STUDENT SOLUTIONS MANUAL
To Accompany

FUNCTIONS MODELING CHANGE
A Preparation for Calculus

PRELIMINARY EDITION

Eric Connally   Deborah Hughes-Hallett
Wellesley College   *Harvard University*

et al.

John Wiley & Sons, Inc.

New York   Chichester   Weinheim   Brisbane   Singapore   Toronto

This project was supported, in part,
by the
National Science Foundation
Opinions expressed are those of the authors
and not necessarily those of the Foundation
Grant No. DUE-9352905

ISBN 0-471-23781-7

Printed in the United States of America

10 9 8 7 6 5 4 3 2

Printed and bound by Courier Kendallville, Inc.

CONTENTS

CHAPTER ONE

Solutions for Section 1.1

1. (a) 69°F
   (b) July 17th and 20th
   (c) Yes. For each date, there is exactly one low temperature.
   (d) No, it is not true that for each low temperature, there is exactly one date: for example, 73° corresponds to both the 17th and 20th.

5. (a) It takes Charles Osgood 60 seconds to read 15 lines, so that means it takes him 4 seconds to read 1 line, 8 seconds for 2 lines, and so on. Table 1.1 shows this. From the table we see that it takes 36 seconds to read 9 lines.

   TABLE 1.1   *The time it takes Charles Osgood to read*

Lines	0	1	2	3	4	5	6	7	8	9	10
Time	0	4	8	12	16	20	24	28	32	36	40

   (b) Figure 1.1 shows the plot of the time in seconds versus the number of lines.

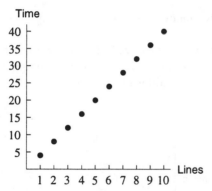

Figure 1.1: The graph of time versus lines

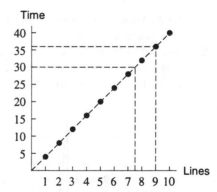

Figure 1.2: The graph of time versus lines

   (c) In Figure 1.2 we have dashed in a line to see the trend. By drawing the vertical line when lines= 9, we see that this corresponds to approximately 36 seconds. By drawing a horizontal line at time= 30 seconds, we see that this corresponds to approximately 7.5 lines.
   (d) If we let T be the time in seconds that it takes to read n lines, then $T = 4n$.

9. Appropriate axes are shown in Figure 1.3.

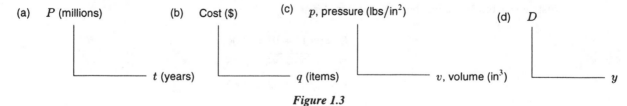

(a) P (millions), t (years) (b) Cost ($), q (items) (c) p, pressure (lbs/in^2), v, volume (in^3) (d) D, y

Figure 1.3

13. Since the tax is $0.06P$, the total cost would be the price of the item plus the tax, or

$$C = P + 0.06P = 1.06P.$$

17. (a) Yes. If the person walks due west and then due north, the distance from home is represented by the hypotenuse of the right triangle that is formed (see Figure 1.4).

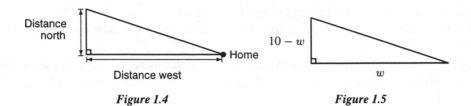

Figure 1.4   *Figure 1.5*

If the distance west is w miles and the total distance walked is 10 miles, then the distance north that she walked is $10 - w$ miles.

We can use the Pythagorean Theorem to find that

$$D = \sqrt{w^2 + (10 - w)^2}.$$

So we know that for each value of w, there corresponds a unique value of D, satisfying the definition of a function.

(b) No. Suppose she walked 10 miles, that is, $x = 10$. She might have walked 1 mile west and 9 miles north, or 2 miles west and 8 miles north, or 3 miles west and 7 miles north, and so on. The right triangles in Fig 1.6 show three different routes she could have taken and still walked 10 miles.

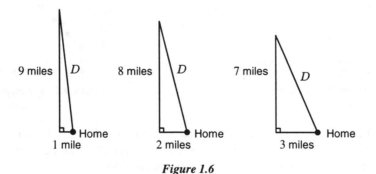

Figure 1.6

Each situation gives a different distance from home. The Pythagorean Theorem shows that the distances from home for these three examples are

$$D = \sqrt{1^2 + 9^2} = 9.06,$$
$$D = \sqrt{2^2 + 8^2} = 8.25,$$
$$D = \sqrt{3^2 + 7^2} = 7.62.$$

Thus, the distance from home cannot be determined from the distance walked.

Solutions for Section 1.2

1. These data are plotted in Figure 1.7. The independent variable is A while the dependent variable is n.

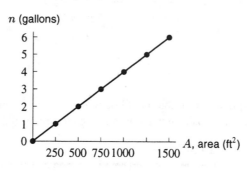

Figure 1.7

5. (a) From Figure 1.8, we see that $P = (b, a)$ and $Q = (d, e)$.

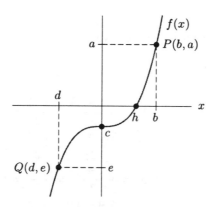

Figure 1.8

(b) To evaluate $f(b)$, we want to find the y-value when the x-value is b. Since (b, a) lies on this graph, we know that the y-value is a, so $f(b) = a$.

(c) To solve $f(x) = e$, we want to find the x-value for a y-value of e. Since (d, e) lies on this curve, $x = d$ is our solution.

(d) To solve $z = f(x)$, we need to first find a value for z; in other words, we need to first solve for $f(z) = c$. Since $(0, c)$ lies on this graph, we know that $z = 0$. Now we need to solve $0 = f(x)$ by finding the point whose y-value is 0. That point is $(h, 0)$, so $x = h$ is our solution.

(e) We know that $f(b) = a$ and $f(d) = e$. Thus, if $f(b) = -f(d)$, we know that $a = -e$.

9. (a) The car's position after 2 hours is denoted by the expression $s(2)$. The position after 2 hours is

$$s(2) = 11(2)^2 + 2 + 100 = 44 + 2 + 100 = 146.$$

(b) This is the same as asking the following question: "For what t is $v(t) = 65$?"

(c)   To find out when the car is going 67 mph, we set $v(t) = 67$. We have

$$22t + 1 = 67$$
$$22t = 66$$
$$t = 3.$$

The car is going 67 mph at $t = 3$, that is, 3 hours after starting. Thus, when $t = 3$, $S(3) = 11(3^2) + 3 + 100 = 202$, so the car's position when it is going 67 mph is 202 miles.

13.   (a)

TABLE 1.2

n	1	2	3	4	5
$s(n)$	1	3	6	10	15

(b)   Substituting into the formula for $s(n)$, we have

$$s(1) = \frac{1(1+1)}{2} = \frac{1 \cdot 2}{2} = 1$$
$$s(2) = \frac{2(2+1)}{2} = \frac{2 \cdot 3}{2} = 3$$
$$s(3) = \frac{3(3+1)}{2} = \frac{3 \cdot 4}{2} = 6$$
$$s(4) = \frac{4(4+1)}{2} = \frac{4 \cdot 5}{2} = 10$$
$$s(5) = \frac{5(5+1)}{2} = \frac{5 \cdot 6}{2} = 15.$$

(c)   To find out how many pins are needed for a 100 row arrangement, we evaluate $s(100)$:

$$s(100) = \frac{100 \cdot 101}{2} = 5050.$$

So 5050 pins are needed.

Solutions for Section 1.3

1. We know that the theater can hold anywhere from 0 to 200 people. Therefore the domain of the function is the integers, n, such that $0 \le n \le 200$.

   We know that each person who enters the theater must pay $4.00. Therefore, the theater makes $(0) \cdot (\$4.00) = 0$ dollars if there is no one in the theater, and $(200) \cdot (\$4.00) = \$800.00$ if the theater is completely filled. Thus the range of the function would be the integers, $4n$, such that $0 \le 4n \le 800$.

   The graph of this function is shown in Figure 1.9.

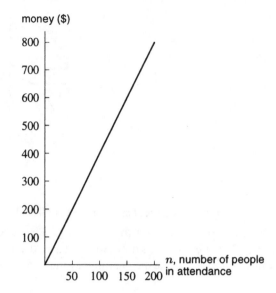

Figure 1.9

5. The graph of $y = \sqrt{8 - x}$ is given in Figure 1.10.

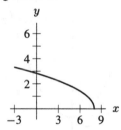

Figure 1.10

The domain is all real $x \le 8$; the range is all $y \ge 0$.

9. The graph of $y = x^2 + 1$ is given in Figure 1.11.

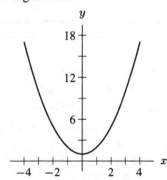

Figure 1.11

The domain is all real x; the range is all $y \ge 1$.

13.   The graph of $y = x^3 + 2$ is given in Figure 1.12.

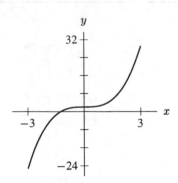

Figure 1.12

The domain is all real x; the range is all real y.

17.   The graph of $y = x^2 - 4$ for $-2 \leq x \leq 3$ is shown in Figure 1.13. From the graph, we see that $y = 0$ at $x = -2$, that y decreases down to -4 at $x = 0$, and then increases to $y = 3^2 - 4 = 5$ at $x = 3$. The minimum value of y is -4, while the maximum value is 5. Thus, on the domain $-2 \leq x \leq 3$, the range is $-4 \leq y \leq 5$.

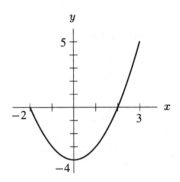

Figure 1.13

Solutions for Section 1.4

1.   (a)   (i)   To evaluate $f(x)$ for $x = 6$, we find the value of $f(x)$ corresponding to an x-value of 6. In this case, the corresponding value is 248. Thus, $f(x)$ at $x = 6$ is 248.

    (ii) $f(5)$ equals the value of $f(x)$ corresponding to $x = 5$, or 145. $f(5) - 3 = 145 - 3 = 142$.

    (iii) $f(5 - 3)$ is the same thing as $f(2)$, which is the value of $f(x)$ corresponding to $x = 2$. Since $f(5 - 3) = f(2)$, and $f(2) = 4$, $f(5 - 3) = 4$.

    (iv) $g(x) + 6$ for $x = 2$ equals $g(2) + 6$. $g(2)$ is the value of $g(x)$ corresponding to an x-value of 2, thus $g(2) = 6$. $g(2) + 6 = 6 + 6 = 12$.

    (v) $g(x + 6)$ for $x = 2$ equals $g(2 + 6) = g(8)$. Looking at the table in the problem, we see that $g(8) = 378$. Thus, $g(x + 6)$ for $x = 2$ equals 378.

    (vi) $g(x)$ for $x = 0$ equals $g(0) = -6$. $3 \cdot (g(0)) = 3 \cdot (-6) = -18$.

    (vii) $f(3x)$ for $x = 2$ equals $f(3 \cdot 2) = f(6)$. From part (a), we know that $f(6) = 248$; thus, $f(3x)$ for $x = 2$ equals 248.

    (viii) $f(x) - f(2)$ for $x = 8$ equals $f(8) - f(2)$. $f(8) = 574$ and $f(2) = 4$, so $f(8) - f(2) = 574 - 4 = 570$.

    (ix) $g(x + 1) - g(x)$ for $x = 1$ equals $g(1 + 1) - g(1) = g(2) - g(1)$. $g(2) = 6$ and $g(1) = -7$, so $g(2) - g(1) = 6 - (-7) = 6 + 7 = 13$.

  (b) (i) To find x such that $g(x) = 6$, we look for the entry in the table at which $g(x) = 6$ and then see what the corresponding x-value is. In this case, it is 2. Thus, $g(x) = 6$ for $x = 2$.

    (ii) We use the same principle as that in part (i): $f(x) = 574$ when $x = 8$.

    (iii) Again, this is just like part (i): $g(x) = 281$ when $x = 7$.

  (c) Solving $x^3 + x^2 + x - 10 = 7x^2 - 8x - 6$ involves finding those values of x for which both sides of the equation are equal, or where $f(x) = g(x)$. Looking at the table, we see that $f(x) = g(x) = -7$ for $x = 1$, and $f(x) = g(x) = 74$ for $x = 4$.

5. (a) To find a point on the graph $k(x)$ with an x-coordinate of -2, we substitute -2 for x in $k(x)$. We obtain $k(-2) = 6 - (-2)^2 = 6 - 4 = 2$. Thus, we have the point $\left(-2, k(-2)\right)$, or $(-2, 2)$.

  (b) To find these points, we want to find all the values of x for which $k(x) = -2$. We have

$$6 - x^2 = -2$$
$$-x^2 = -8$$
$$x^2 = 8$$
$$x = \pm 2\sqrt{2}.$$

    Thus, the points $(2\sqrt{2}, -2)$ and $(-2\sqrt{2}, -2)$ both have a y-coordinate of -2.

  (c) Figure 1.14 shows the desired graph.

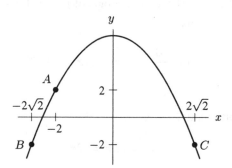

Figure 1.14

    The point in part (a) is $(-2, 2)$. We have called this point A on the graph in Figure 1.14. There are two points in part (b): $(-2\sqrt{2}, -2)$ and $(2\sqrt{2}, -2)$. We have called these points B and C, respectively, on the graph in Figure 1.14.

  (d) For $p = 2$, $k(p) - k(p-1) = k(2) - k(1)$. $k(2) = 6 - 2^2 = 6 - 4 = 2$, while $k(1) = 6 - (1)^2 = 6 - 1 = 5$. Thus, $k(2) - k(1) = 2 - 5 = -3$.

9. (a) (ii) The \$5 tip is added to the fare $f(x)$, so the total is $f(x) + 5$.

  (b) (iv) There were 5 extra miles so the trip was $x + 5$. I paid $f(x + 5)$.

  (c) (i) Each trip cost $f(x)$ and I paid for 5 of them, or $5f(x)$.

  (d) (iii) The miles were 5 times the usual so $5x$ is the distance, and the cost is $f(5x)$.

13.   (a)   The formula is $A = f(r) = \pi r^2$.
      (b)   If the radius is increased by 10%, then the new radius is $r + (10\%)r = (110\%)r = 1.1r$. We want to know the output when our input is $1.1r$, so the appropriate expression is $f(1.1r)$.
      (c)   Since $f(1.1r) = \pi(1.1r)^2 = 1.21\pi r^2$, the new area is the old area multiplied by 1.21, or 121% of the old area. In other words, the area of a circle is increased by 21% when its radius is increased by 10%.

Solutions for Section 1.5

1.   (a)   Since $V = (4\pi/3)r^3$, the constant of proportionality is $k = 4\pi/3$.
     (b)   Since $V = (4\pi/3)r^3$, we have $p = 3$.
     (c)   The graph is shown in Figure 1.15.

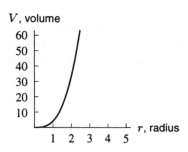

Figure 1.15

Notice that it has the same shape as the graph of $y = x^3$ for $x \geq 0$.

5.   This is a case of direct proportionality.

$$y = \frac{(0.34)}{2}x = 0.17x = (0.17)x^{(1)}.$$

Thus $k = 0.17$ and $p = 1$.

9.   This is a case of direct proportionality.

$$C = 2\pi r = (2\pi)r^{(1)}.$$

Thus $k = 2\pi$ and $p = 1$.

13.   This is a power function.

$$y = (2x)^5 = (2^5)x^5 = (32)x^{(5)}.$$

Thus, $k = 32$ and $p = 5$.

17.   This is a power function.

$$l = \frac{-5}{t^{-4}} = (-5)t^{(4)}.$$

Thus $k = -5$ and $p = 4$.

21.   This is not a power function because we cannot get it into the form $Q = kx^p$ for any Q or x. Instead, we have a function of the form $Q = kp^x$.

25. (a) Table 1.3 shows the circulation times in seconds for various mammals.

TABLE 1.3

Animal	Body mass (kg)	Circulation time (sec)
Blue whale	91000	302
African elephant	5450	150
White rhinoceros	3000	129
Hippopotamus	2520	123
Black rhinoceros	1170	102
Horse	700	90
Lion	180	64
Human	70	50

(b) If a mammal of mass m has a circulation time of T, then

$$T = 17.4m^{1/4}.$$

If a mammal of mass M has twice the circulation time, then

$$2T = 17.4M^{1/4}.$$

We want to find the relationship between m and M, so we divide these two equations, giving

$$\frac{2T}{T} = \frac{17.4M^{1/4}}{17.4m^{1/4}}.$$

Simplifying, we have

$$2 = \frac{M^{1/4}}{m^{1/4}}.$$

Taking the fourth power of both sides, we get

$$2^4 = \frac{M}{m},$$

and thus

$$16 = \frac{M}{m}.$$

The body mass of the animal with the larger circulation time is 16 times the body mass of the other animal.

29. Since $A = \pi r^2$ and $r = d/2$, substituting for r in the formula for A gives

$$A = \pi \left(\frac{d}{2}\right)^2 = \frac{\pi d^2}{4}.$$

33. $f(x) = \begin{cases} -1, & -1 \leq x < 0 \\ 0, & 0 \leq x < 1 \\ 1, & 1 \leq x < 2 \end{cases}$ is shown in Figure 1.16.

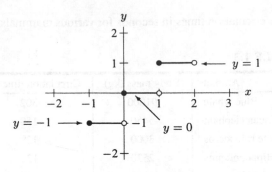

Figure 1.16

37. (a) Up to 1/8 mile, the cost is $1.50. The next 1/8 mile, (up to 2/8 mile) adds $0.25, giving a fare of $1.75. For a journey of 3/8 mile, another $0.25 is added for a fare of $2.00. Each additional 1/8 mile gives an another increment of $0.25. See Table 1.4.

TABLE 1.4

Miles	0	1/8	2/8	3/8	4/8	5/8	6/8	7/8	1
Cost	0	1.50	1.75	2.00	2.25	2.50	2.75	3.00	3.25

(b) The table shows that the cost for a 5/8 mile trip is $2.50.

(c) From the table, the maximum distance one can travel for $3.00 is 7/8 mile.

(d)

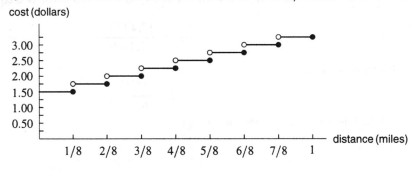

Figure 1.17

Solutions for Section 1.6

1. (a) Let $s = C(t)$ be the sales (in millions) of CDs in year t. Then

$$\text{Average rate of change of } s \text{ from } t = 1982 \text{ to } t = 1983 = \frac{\Delta s}{\Delta t} = \frac{C(1983) - C(1982)}{1983 - 1982}$$

$$= \frac{0.8 - 0}{1}$$

$$= 0.8 \text{ million discs/year.}$$

Let $q = L(t)$ be the sales (in millions) of LPs in year t. Then

$$\begin{aligned}\text{average rate of change of } q \\ \text{from } t = 1982 \text{ to } t = 1983\end{aligned} = \frac{\Delta q}{\Delta t} = \frac{L(1983) - L(1982)}{1983 - 1982}$$

$$= \frac{210 - 244}{1}$$

$$= -34 \text{ million records/year.}$$

(b) By the same argument

$$\begin{aligned}\text{average rate of change of } s \\ \text{from } t = 1986 \text{ to } t = 1987\end{aligned} = \frac{\Delta s}{\Delta t} = \frac{C(1987) - C(1986)}{1987 - 1986}$$

$$= \frac{102 - 53}{1}$$

$$= 49 \text{ million discs/year.}$$

$$\begin{aligned}\text{average rate of change of } q \\ \text{from } t = 1986 \text{ to } t = 1987\end{aligned} = \frac{\Delta q}{\Delta t} = \frac{L(1987) - L(1986)}{1987 - 1986}$$

$$= \frac{105 - 125}{1}$$

$$= -20 \text{ million records/year.}$$

(c) The fact that $\Delta s / \Delta t = 0.8$ tells us that CD sales increased at an average rate of 0.8 million discs/year during 1982. The fact that $\Delta s / \Delta t = 49$ tells us that CD sales increased at an average rate of 49 million discs/year during 1986.

The fact that $\Delta q / \Delta t = -34$ means that LP sales decreased at an average rate of 34 million records/year during 1982. The fact that the average rate of change is negative tells us that annual sales are decreasing.

The fact that $\Delta q / \Delta t = -20$ means that LP sales decreased at an average rate of 20 million records/year during 1986.

5. Starting on the left, we see that the function is increasing until approximately $x = -1.5$. It then decreases until approximately $x = 0$. Then it increases until approximately $x = 1.5$. After that the function decreases. Thus, $y = f(x)$ is increasing approximately on the intervals $x < -1.5$, $0 < x < 1.5$, and decreasing on the intervals $-1.5 < x < 0$, $x > 1.5$.

As for concavity, starting at the left we see that the function bends downward until approximately $x = -1$. From $x = -1$ to $x = 1$, it bends upward, and after $x = 1$ it bends downward again. Therefore, $y = f(x)$ appears to be concave down on the intervals $x < -1$ and $x > 1$ and concave up on the interval $-1 < x < 1$.

9. (a) For 1980 to 1990, the rate of change of P_1 is

$$\frac{\Delta P_1}{\Delta t} = \frac{83 - 53}{1990 - 1980} = \frac{30}{10} = 3 \text{ hundred people per year,}$$

while for P_2 we have

$$\frac{\Delta P_2}{\Delta t} = \frac{70 - 85}{1990 - 1980} = \frac{-15}{10} = \frac{-3}{2} \text{ hundred people per year.}$$

(b) For 1985 to 1997,

$$\frac{\Delta P_1}{\Delta t} = \frac{93 - 73}{1997 - 1985} = \frac{20}{12} = \frac{5}{3} \text{ hundred people per year,}$$

and

$$\frac{\Delta P_2}{\Delta t} = \frac{65 - 75}{1997 - 1985} = \frac{-10}{12} = \frac{-5}{6} \text{ hundred people per year.}$$

(c) For 1980 to 1997,

$$\frac{\Delta P_1}{\Delta t} = \frac{93 - 53}{1997 - 1980} = \frac{40}{17} \text{ hundred people per year,}$$

and

$$\frac{\Delta P_2}{\Delta t} = \frac{65 - 85}{1997 - 1980} = \frac{-20}{17} \text{ hundred people per year.}$$

13. (a) Since Δt refers to the change in the numbers of years, we calculate

$$\Delta t = 1965 - 1960 = 5, \qquad \Delta t = 1970 - 1965 = 5, \qquad \text{and so on.}$$

Since the entries in the table are all 5 years apart, we see that $\Delta t = 5$ for all consecutive entries.

(b) Since ΔG is the change in the amount of garbage produced, for the period 1960-1965 we have

$$\Delta G = 105 - 90 = 15.$$

Continuing in this way gives the following:

TABLE 1.5

time period	1960-65	1965-70	1970-75	1975-80	1980-85	1985-90
ΔG	15	15	10	20	15	15

(c) Not all of the ΔG values are the same. We know that all the values of Δt are the same. If we knew that all the values of ΔG were the same, we could say that $\Delta G/\Delta t$, the average amount of garbage we produce each year, was constant. Since, on the contrary, ΔG is not constant, we conclude that $\Delta G/\Delta t$ is not constant. This tells us that the amount of garbage being produced is changing, but not at a constant rate.

17. (a) This describes a situation in which y is increasing rapidly at first, then very slowly at the end. In Table (E), y increases dramatically at first (from 20 to 275) but is hardly growing at all by the end. In Graph (I), y is increasing at a constant rate, while in Graph (II), it is increasing faster at the end. Thus, scenario (a) matches with Table (E) and Graph (III).

(b) Here, y is growing at a constant rate. In Table (G), y increases by 75 units for every 5-unit increase in x. A constant increase in y relative to x means a straight line, that is, a line with a constant slope. This is found in Graph (I).

(c) In this scenario, y is growing at a faster and faster rate as x gets larger. In Table (F), y starts out by growing by 16 units, then 30, then 54, and so on, so Table (F) refers to this case. In Graph (II), y is increasing faster and faster as x gets larger.

21. (a) A table of values for $y = 1/x^2$ follows:

x	-5	-4	-3	-2	-1	$-\frac{3}{4}$	$-\frac{1}{2}$	$-\frac{1}{4}$
y	$\frac{1}{25}$	$\frac{1}{16}$	$\frac{1}{9}$	$\frac{1}{4}$	1	$\frac{16}{9}$	4	16

x	0	$\frac{1}{4}$	$\frac{1}{2}$	$\frac{3}{4}$	1	2	3	4	5
y	undefined	16	4	$\frac{16}{9}$	1	$\frac{1}{4}$	$\frac{1}{9}$	$\frac{1}{16}$	$\frac{1}{25}$

(b) The y-values in the table range from near 0 to 16.

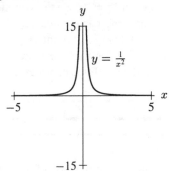

Figure 1.18

(c) Domain is all real numbers except 0.
Range is $0 < y < \infty$.

(d) Increasing: $-\infty < x < 0$.
Decreasing: $0 < x < \infty$.

(e) Concave up: $-\infty < x < 0$ and $0 < x < \infty$.

25. (a)

x	-3	-2.5	-2.25	-2.1	-2.01	-2	-1.99	-1.9	-1.75	-1.5	-1
$g(x)$	1	4	16	100	$10{,}000$	undefined	$10{,}000$	100	16	4	1

As x approaches -2 from the left the function takes on very large positive values. As x approaches -2 from the right the function takes on very large positive values.

(b)

x	5	10	100	1000
$g(x)$	0.02	0.007	$9.6 \cdot 10^{-5}$	10^{-6}

x	-5	-10	-100	-1000
$g(x)$	0.111	0.016	10^{-4}	10^{-6}

For $x > -2$, as x increases, $f(x)$ approaches 0 from above. For $x < -2$ as x decreases, $f(x)$ approaches 0 from above.

(c) The horizontal asymptote is $y = 0$ (the x-axis). The vertical asymptote is $x = -2$. See Figure 1.19.

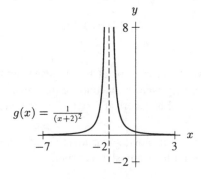

Figure 1.19

29.   (a)   From O to A, the rate is zero, so no water is flowing into the reservoir, and the volume remains constant. From A to B, the rate is increasing, so the volume is going up more and more quickly. From B to C, the rate is holding steady, but water is still going into the reservoir—it's just going in at a constant rate. So volume is increasing on the interval from B to C. Similarly, it is increasing on the intervals from C to D and from D to E. Even on the interval from E to F, water is flowing into the reservoir; it is just going in more and more slowly (the *rate* of flow is decreasing, but the total amount of water is still increasing). So we can say that the volume of water increases throughout the interval from A to F.
       (b)   The volume of water is constant when the rate is zero, that is from O to A.
       (c)   According to the graph, the rate at which the water is entering the reservoir reaches its highest value at $t = D$ and stays at that high value until $t = E$. So the volume of water is increasing most rapidly from D to E. (Be careful. The rate itself is increasing most rapidly from C to D, but the volume of water is increasing fastest when the rate is at its highest points.)
       (d)   When the rate is negative, water is leaving the reservoir, so its volume is decreasing. Since the rate is negative from F to I, we know that the volume of water *decreases* on that interval.

Solutions for Chapter 1 Review

1.   (a)   Using the vertical line test, we can see that y is not a function of x.
      (b)   To determine whether x is a function of y, we want to know if, for each value of y, there is a unique value of x associated with it. If we were to draw a horizontal line through the graph, representing one value of y, we could see that the line intersects the graph in more than one place. This tells us that there are many values of x corresponding to a value y, so this graph does not define a function.
      (c)   If there were an interval on the x-axis for which y is a function of x, then there would be an interval for which each value of x would pass the vertical line test. The only place on the graph where that happens is in the interval shown in Figure 1.20.

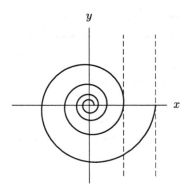

Figure 1.20

5.   (a)   At the end of the race, Owens was running at 12 yards/sec and the horse was running at 20 yards/sec.
      (b)   We can find the time when they were both running the same speed by finding the point on the graph when the two lines intersect. This occurs at $t = 6$. So they were both running the same speed after 6 seconds.

9.   Figure 1.21 shows a possible graph of blood sugar level as a function of time over one day. Note that the actual curve is smooth, and does not have any sharp corners.

blood-sugar level

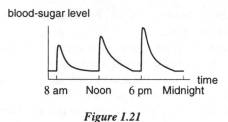

Figure 1.21

13. (a) If $V = \pi r^2 h$ and $V = 355$, then $\pi r^2 h = 355$. So $h = (355)/(\pi r^2)$. Thus, since

$$A = 2\pi r^2 + 2\pi rh,$$

we have

$$A = 2\pi r^2 + 2\pi r \left(\frac{355}{\pi r^2}\right),$$

and

$$A = 2\pi r^2 + \frac{710}{r}.$$

(b)

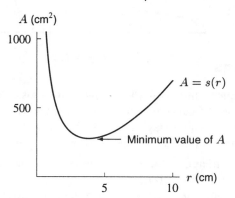

Figure 1.22: Graph of $A(r)$ for $0 < r \le 10$

(c) The domain is any positive value or $r > 0$, because (in practice) a cola can could have as large a radius as you wanted (it would just have to be very short to maintain its 12 oz size). From the graph in (b), the value of A is never less than about 277.5 cm². Thus, the range is $A > 277.5$ cm² (approximately).

(d) They need a little more than 277.5 cm² per can. The minimum A-value occurs (from graph) at $r \approx 3.83$ cm, and since $h = 355/\pi r^2$, $h \approx 7.7$ cm.

(e) Since the radius of a real cola can is less than the value required for the minimum value of A, it must use more aluminum than necessary. This is because the minimum value of A has $r \approx 3.83$ cm and $h \approx 7.7$ cm. Such a can has a diameter of $2r$ or 7.66 cm. This is roughly equal to its height—holding such a can would be difficult. Thus, real cans are made with slightly different dimensions.

17. (a) $g(100) = 100\sqrt{100} + 100 \cdot 100 = 100 \cdot 10 + 100 \cdot 100 = 11,000$
     (b) $g(4/25) = 4/25 \cdot \sqrt{4/25} + 100 \cdot 4/25 = 4/25 \cdot 2/5 + 16 = 8/125 + 16 = 16.064$
     (c) $g(1.21 \cdot 10^4) = g(12100) = (12100)\sqrt{12100} + 100 \cdot (12100) = 2,541,000$

21. (a) Since for any value of x that you might choose you can find a corresponding value of $m(x)$, we can say that the domain of $m(x) = 9 - x$ is all real numbers.
     For any value of $m(x)$ there is a corresponding value of x. So the range is also all real numbers.

(b) Since you can choose any value of x and find an associated value for $n(x)$, we know that the domain of this function is all real numbers.

However, there are some restrictions on the range. Since x^4 is always positive for any value of x, $9 - x^4$ will have a largest value of 9 when $x = 0$. So the range is $n(x) \leq 9$.

(c) The expression $x^2 - 9$, found inside the square root sign, must always be non-negative. This happens when $x \geq 3$ or $x \leq -3$, so our domain is $x \geq 3$ or $x \leq -3$.

For the range, the smallest value $\sqrt{x^2 - 9}$ can have is zero. There is no largest value, so the range is $q(x) \geq 0$.

25. (a) For the narrow parts of the river, the raft (solid line) will be traveling faster than the raft travels in the wide parts. This means that on the narrow parts it will take less time for the raft to travel the same distance, so the curve will be steeper in those parts.

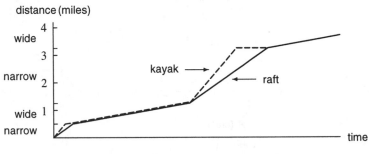

Figure 1.23

(b) In the narrow regions, the kayak (dotted line) moves faster than the raft. So at any time when they are in a narrow region, the kayak will be farther along the river, and thus "higher-up" on the graph. Where the river is wide, the kayak stays alongside the raft so their curves coincide along those regions. The horizontal segments represent the time spent by the kayak waiting for the raft to catch up.

29. (a) The smaller the difference, the smaller the refund. The smallest possible difference is \$0.01. This translates into a refund of \$1.00 + \$0.01 = \$1.01.

(b) Looking at the refund rules, we see that there are three separate cases to consider. The first case is when 10 times the difference is less than \$1. If the difference is more than 0 but less than 10¢, and you will receive \$1 plus the difference. The formula for this is:

$$y = 1 + x \quad \text{for} \quad 0 < x < 0.10.$$

In the second case, 10 times the difference is between \$1 and \$5. This will be true if the difference is between 10¢ and 50¢. The formula for this is:

$$y = 10x + x \quad \text{for} \quad 0.10 \leq x \leq 0.50.$$

In the third case, 10 times the difference is more than \$5. If the difference is more than 50¢, then you receive \$5 plus the difference or:

$$y = 5 + x \quad \text{for} \quad x > 0.50.$$

Putting these cases together, we get:

$$y = \begin{cases} 1 + x & \text{for } 0 < x < 0.1 \\ 10x + x & \text{for } 0.1 \leq x \leq 0.5 \\ 5 + x & \text{for } x > 0.5. \end{cases}$$

(c)   We want x such that $y = 9$. Since the highest possible value of y for the first case occurs when $x = 0.09$, and $y = 1 + 0.09 = \$1.09$, the range for this case does not go high enough. The highest possible value for the second case occurs when $x = 0.5$, and $y = 10(0.5) + 0.5 = \$5.50$. This range is also not high enough. So we look to the third case where $x > 0.5$ and $y = 5 + x$. Solving $5 + x = 9$ we find $x = 4$. So the price difference would have to be $4.

(d)

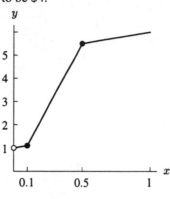

Figure 1.24

CHAPTER TWO

Solutions for Section 2.1

1. The function f could be linear because the value of x increases by $\Delta x = 5$ each time and $f(x)$ increases by $\Delta f(x) = 10$ each time. Assuming that any values of f not shown by the table follow this same pattern, the function f is linear.

   The function g is not linear even though $g(x)$ increases by $\Delta g(x) = 50$ each time. This is because the value of x does not increase by the same amount each time. The value of x increases from 0 to 100 to 300 to 600 taking steps that get larger each time.

   Similarly, the function h is not linear either even though the value of x increases by $\Delta x = 10$ each time. This is because $h(x)$ does not increase by the same amount each time. The value of $h(x)$ increases from 20 to 40 to 50 to 55 taking smaller steps each time.

   The function j could be linear because if the pattern continues for values of $j(x)$ that are not shown, we see that a one unit increase in x corresponds to a constant decrease of two units in y.

5. We know that the area of a circle of radius r is

$$\text{Area} = \pi r^2$$

while its circumference is given by

$$\text{Circumference} = 2\pi r.$$

Thus, a table of values for area and circumference is

TABLE 2.1

Radius	0	1	2	3	4	5	6
Area	0	π	4π	9π	16π	25π	36π
Circumference	0	2π	4π	6π	8π	10π	12π

(a) In the area function we see that the rate of change between pairs of points does not remain constant and thus the function is not linear. For example, the rate of change between the points $(0,0)$ and $(2,4\pi)$ is not equal to the rate of change between the points $(3,9\pi)$ and $(6,36\pi)$. The rate of change between $(0,0)$ and $(2,4\pi)$ is

$$\frac{\Delta\text{area}}{\Delta\text{radius}} = \frac{4\pi - 0}{2 - 0} = \frac{4\pi}{2} = 2\pi$$

while the rate of change between $(3,9\pi)$ and $(6,36\pi)$ is

$$\frac{\Delta\text{area}}{\Delta\text{radius}} = \frac{36\pi - 9\pi}{6 - 3} = \frac{27\pi}{3} = 9\pi.$$

On the other hand, if we take only pairs of points from the circumference function, we see that the rate of change remains constant. For instance, for the pair $(0,0)$, $(1,2\pi)$ the rate of change is

$$\frac{\Delta\text{circumference}}{\Delta\text{radius}} = \frac{2\pi - 0}{1 - 0} = \frac{2\pi}{1} = 2\pi.$$

For the pair $(2, 4\pi)$, $(4, 8\pi)$ the rate of change is

$$\frac{\Delta \text{circumference}}{\Delta \text{radius}} = \frac{8\pi - 4\pi}{4 - 2} = \frac{4\pi}{2} = 2\pi.$$

For the pair $(1, 2\pi)$, $(6, 12\pi)$ the rate of change is

$$\frac{\Delta \text{circumference}}{\Delta \text{radius}} = \frac{12\pi - 2\pi}{6 - 1} = \frac{10\pi}{5} = 2\pi.$$

Picking any pair of data points would give a rate of change of 2π.
(b) The graphs for area and circumference as indicated in Table 2.1 are shown in Figure 2.1 and Figure 2.2.

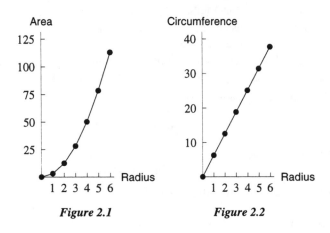

Figure 2.1   Figure 2.2

(c) From part (a) we see that the rate of change of the circumference function is 2π. This tells us that for a given circle, when we increase the length of the radius by one unit, the length of the circumference would increase by 2π units. Equivalently, if we decreased the length of the radius by one unit, the length of the circumference would decrease by 2π.

9. The following table shows the population of the team as a function of the number of years since 1978.

TABLE 2.2

t	P
0	18,310
1	$18,310 + 58$
2	$18,310 + 58 + 58 = 18,310 + 2 \times 58$
3	$18,310 + 3 \times 58$
4	$18,310 + 4 \times 58$
...	
t	$18,310 + t \times 58$

So, a formula is $P = 18,310 + 58t$.

13. (a) $F = 2C + 30$
     (b) If $C = -5$, then the approximate Fahrenheit temperature is $2(-5) + 30 = -10 + 30 = 20$ degrees and the actual temperature is $9/5(-5) + 32 = -9 + 32 = 23$ degrees. Their difference is $20 - 23 = -3$, so the approximation is 3 degrees too low in this case.

     If $C = 30°$, then the approximate temperature is $2(30) + 30 = 90$ which the actual temperature is $(9/5)(30) + 32 = 54 + 32 = 86$. Here, the difference is $90 - 86 = 4$, so the approximate temperature is 4 degrees above the actual.

     Since we are finding the difference for a number of values, it would perhaps be easier to find a formula for the difference:

     $$\text{Difference} = \text{Approximate value} - \text{Actual value}$$
     $$= (2C + 30) - \left(\frac{9}{5}C + 32\right) = \frac{1}{5}C - 2.$$

     If the Celsius temperature is $-5°$, $(1/5)C - 2 = (1/5)(-5) - 2 = -1 - 2 = -3$. This agrees with our results above.

     Similarly, we see that when $C = 0$, the difference is $(1/5)(0) - 2 = -2$ or 2 degrees too low. When $C = 15$, the difference is $(1/5)(15) - 2 = 3 - 2 = 1$ or 1 degree too high. When $C = 30$, the difference is $(1/5)(30) - 2 = 6 - 2 = 4$ or 4 degrees too high.
     (c) We are looking for a temperature C, for which the difference between the approximation and the actual formula is zero.

     $$\frac{1}{5}C - 2 = 0$$
     $$\frac{1}{5}C = 2$$
     $$C = 10$$

     Another way we can solve for a temperature C is to equate our approximation and the actual value.

     $$\text{Approximation} = \text{Actual value}$$
     $$2C + 30 = 1.8C + 32,$$
     $$0.2C = 2$$
     $$C = 10$$

     So the approximation agrees with the actual formula at $10°$ Celsius.

17. Any function will look linear if viewed in a small enough window. This function is not linear. We see this by graphing the function in the larger window $-100 \leq x \leq 100$.

21. Since the radius is 10 miles, the longest ride will not be more than 20 miles. The maximum cost will therefore occur when $d = 20$, so the maximum cost is $C = 1.50 + 2d = 1.50 + 2(20) = 1.50 + 40 = 41.50$. Therefore, the window should be at least $0 \leq d \leq 20$ and $0 \leq C \leq 41.50$.

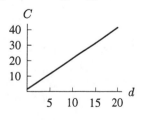

Figure 2.3

Solutions for Section 2.2

1. (a)

$$3x + 5y = 20$$
$$5y = 20 - 3x$$
$$y = \frac{20}{5} - \frac{3x}{5}$$
$$y = 4 - \frac{3}{5}x$$

(b)

$$0.1y + x = 18$$
$$0.1y = 18 - x$$
$$y = \frac{18}{0.1} - \frac{x}{0.1}$$
$$y = 180 - 10x$$

(c)

$$y - 0.7 = 5(x - 0.2)$$
$$y - 0.7 = 5x - 1$$
$$y = 5x - 1 + 0.7$$
$$y = 5x - 0.3$$
$$y = -0.3 + 5x$$

(d)

$$5(x + y) = 4$$
$$5x + 5y = 4$$
$$5y = 4 - 5x$$
$$\frac{5y}{5} = \frac{4}{5} - \frac{5x}{5}$$
$$y = \frac{4}{5} - x$$

(e)

$$5x - 3y + 2 = 0$$
$$-3y = -2 - 5x$$
$$y = \frac{-2}{-3} - \frac{5}{-3}x$$
$$y = \frac{2}{3} + \frac{5}{3}x$$

(f)

$$\frac{x+y}{7} = 3$$
$$x + y = 21$$
$$y = 21 - x$$

(g)

$$3x + 2y + 40 = x - y$$
$$2y + y = x - 3x - 40$$
$$3y = -40 - 2x$$
$$y = -\frac{40}{3} - \frac{2}{3}x$$

(h)   Not possible, the slope is not defined (vertical line).

5.   We know that our function is linear so it is of the form

$$f(t) = b + mt.$$

To solve this problem we can choose any two points to find the slope. We will use the points $(1.4, 0.492)$ and $(1.5, 0.37)$. Now

$$m = \frac{\Delta f}{\Delta t}$$
$$= \frac{0.37 - 0.492}{1.5 - 1.4}$$
$$= \frac{-0.122}{0.1} = -1.22.$$

Thus $f(t)$ is of the form $f(t) = b - 1.22t$. Substituting the coordinates of the point $(1.5, 0.37)$ we get

$$0.37 = b - 1.22 \cdot 1.5.$$

In other words,

$$b = 0.37 - (-1.22 \cdot 1.5) = 0.37 + 1.83 = 2.2.$$

Thus

$$f(t) = 2.2 - 1.22t.$$

9.   (a)   A table of the allowable combinations of sesame and poppy seed rolls is shown below.

TABLE 2.3

s, sesame seed rolls	0	1	2	3	4	5	6	7	8	9	10	11	12
p poppy seed rolls	12	11	10	9	8	7	6	5	4	3	2	1	0

(b)   The sum of s and p is 12. So we can write $s + p = 12$, or $p = 12 - s$.

(c)

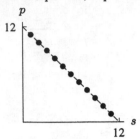

Figure 2.4

13.   Point P is on the curve $y = x^2$ and so its coordinates are $(2, 2^2) = (2, 4)$. Since line l contains point P and has slope 4, its equation is

$$y = b + mx.$$

Using $P = (2, 4)$ and $m = 4$, we get

$$4 = b + 4(2)$$
$$4 = b + 8$$
$$-4 = b$$

so,

$$y = -4 + 4x.$$

17.

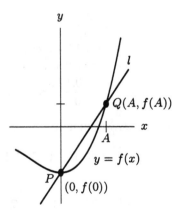

Figure 2.5

Using points $P = (0, f(0))$ and $Q = (A, f(A))$, we can find the slope of the line to be

$$m = \frac{\Delta y}{\Delta x} = \frac{f(A) - f(0)}{A - 0} = \frac{f(A) - f(0)}{A}.$$

Since the y-intercept of l is $b = f(0)$, we have

$$y = f(0) + \frac{f(A) - f(0)}{A} x.$$

21. (a) If c increases you can buy more. If you buy only apples, the increase will mean buying more apples, so the x-intercept increases. Likewise, if you buy only bananas, the y-intercept will show an increase. In Figure 2.17, only l_1 has increases in both the x and y–intercepts. An alternate approach is to note the slope of l, found to be $-$apple price/banana price, in Problem 19, will not change if the prices do not change. We see l_1 is the only line that has the same slope as l.

    (b) If the price of apples goes up and the budget is the same, then when buying only apples, we would get fewer apples. Graphically, this means the x-intercept will decrease in comparison to the x-intercept of line l. We see this is only true for line l_3.

    (c) With a budget increase, while the price of bananas stays fixed, the y-intercept increases from the original budget line. Line l_2 is a likely suspect. We confirm this by looking at the slope of the original line, $-$apple price/banana price. Since the apple price increases while the banana price stays fixed, this slope must become more negative. Notice line l_2 has a more negative slope than the original budget line.

    (d) We have used the three choices and suspect that there is no match, but we should not be too hasty. Perhaps line l_1 or l_2 could match because we have no scales shown. However, if we consider the slope of l, we see then a decrease in the price of apples with the banana price the same will mean a less negative slope than line l. There are no such choices.

Solutions for Section 2.3

1. The functions f and g have the same y-intercept, $b = 20$. u and v both have y-intercept $b = 60$. f and g are increasing functions, with slopes $m = 2$ and $m = 4$, respectively. u and v are decreasing functions, with slopes $m = -1$ and $m = -2$, respectively.

    The figure shows that graphs A and B describe increasing functions with the same y-intercept. The functions f and g are good candidates since they are both linear functions with positive slope and their y-intercepts coincide. Since graph A is steeper than graph B, the slope of A is greater than the slope of B. The slope of g is larger than the slope of f, so graph A corresponds to g and graph B corresponds to f.

    Graphs D and E describe decreasing functions with the same y-intercept. u and v are good candidates since they both have negative slope and their y-intercepts coincide. Graph E is steeper than graph D. Thus, graph D corresponds to u, and graph E to v. Note that graphs D and E start at a higher point on the y-axis than A and B do. This corresponds to the fact that the y-intercept $b = 60$ of u and v is above the y-intercept $b = 20$ of f and g.

    This leaves graph C and the function h. The y-intercept of h is -30, corresponding to the fact that graph C starts below the x-axis. The slope of h is 2, the same slope as f. Since graph C appears to climb at the same rate as graph B, it seems reasonable that f and h should have the same slope.

5. (a) $f(x)$ has a y-intercept of 1 and a positive slope. Thus, Figure 2.32 (ii) must be the graph of $f(x)$.

    (b) $g(x)$ has a y-intercept of 1 and a negative slope. Thus, Figure 2.32 (iii) must be the graph of $g(x)$.

    (c) $h(x)$ is a constant function with a y intercept of 1. Thus, Figure 2.32 (i) must be the graph of $h(x)$.

9. (a) See Figure 2.6.

    (b) See Figure 2.7.

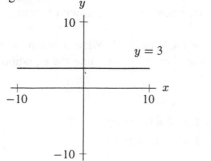

Figure 2.6

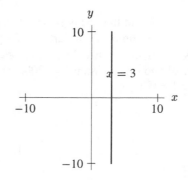

Figure 2.7

(c)   Yes, $y = 3 + 0x$.

(d)   No, since the slope is undefined, and there is no y-intercept.

13.   Since P is the x-intercept, we know that point P has y-coordinate $= 0$, and if the x-coordinate is x_0, we can calculate the slope of line l using $P(x_0, 0)$ and the other given point $(0, -2)$.

$$m = \frac{-2 - 0}{0 - x_0} = \frac{-2}{-x_0} = \frac{2}{x_0}.$$

We know this equals 2, since l is parallel to $y = 2x + 1$ and therefore must have the same slope. Thus we have

$$\frac{2}{x_0} = 2.$$

So $x_0 = 1$ and the coordinates of P are $(1, 0)$.

17.

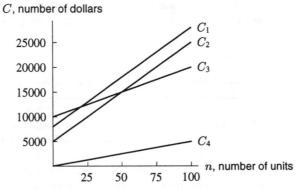

21.   (a)   We are looking at the amount of municipal solid waste, W, as a function of year, t, and the two points are $(1960, 82.3)$ and $(1980, 139.1)$. For the model, we assume that the quantity of solid waste is a linear function of year. The slope of the line is

$$m = \frac{139.1 - 82.3}{1980 - 1960} = \frac{56.8}{20} = 2.84 \frac{\text{millions of tons}}{\text{year}}.$$

This slope tells us that the amount of solid waste generated in the cities of the US has been going up at a rate of 2.84 million tons per year. To find the equation of the line, we must find the vertical intercept. We substitute the point $(1960, 82.3)$ and the slope $m = 2.84$ into the equation $W = b + mt$:

$$W = b + mt$$
$$82.3 = b + (2.84)(1960)$$
$$82.3 = b + 5566.4$$
$$-5484.1 = b.$$

The equation of the line is $W = -5484.1 + 2.84t$, where W is the amount of municipal solid waste in the US in millions of tons, and t is the year.

(b)   How much solid waste does this model predict in the year 2000? We can graph the line and find the vertical coordinate when $t = 2000$, or we can substitute $t = 2000$ into the equation of the line, and solve for W:

$$W = -5484.1 + 2.84t$$
$$W = -5484.1 + (2.84)(2000)$$
$$W = -5484.1 + 5680 = 195.9.$$

The model predicts that in the year 2000, the solid waste generated by cities in the US will be 195.9 million tons.

Solutions for Section 2.4

1. (a)

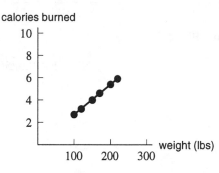

Figure 2.8

    (b) See Figure 2.8.
    (c) Since the points all seem to be in a linear alignment, the correlation coefficient is close to one.

5. (a)

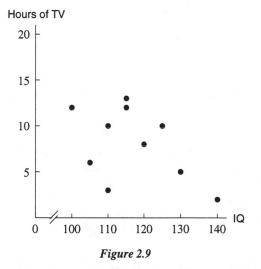

Figure 2.9

    (b) The scatterplot suggests that as IQ increases, the number of hours of TV viewing decreases. The points, though, are not close to being on a line, so $r \approx -1/2$.
    (c)

$$y = 27.5139 - 0.1674x$$
$$r = -0.5389$$

Solutions for Chapter 2 Review

1. (a) From the table we find that a 200lb person uses 5.4 calories per minute while walking. So a half-hour, or a 30 minute, walk would burn $30(5.4) = 162$ calories.

(b)   The number of calories used per minute is approximately proportional to the person's weight. The relationship is an approximately linear increasing function, where weight is the independent variable and number of calories burned is the dependent variable.

(c)   (i)   Since the function is approximately linear, its equation is $y = b + mx$. The slope is

$$m = \frac{3.2 - 2.7}{120 - 100} = \frac{0.5}{20} = 0.025 \text{ cal/lb.}$$

Using the point $(100, 2.7)$ we have

$$2.7 = b + 0.025(100)$$
$$b = 0.2.$$

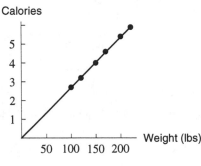

Figure 2.10

Examining our graph, we find that it nearly passes through the origin and has a slope of $2.5/100$. Thus,

$$\text{Calories} = 0.025 \text{ weight.}$$

(ii)   The intercept $(0, 0.2)$ is the number of calories burned by a weightless runner. This implies that most of the calories burned are due to moving your weight.

(iii)   Domain $0 < w$; range $0 < c$

(iv)   Evaluating our function at 135,

$$\text{Calories} = 0.025(135) \approx 3.4.$$

5.   We would like to find a table value that corresponds to $n = 0$. The pattern from the table, is that for each decrease of 25 in n, $C(n)$ goes down by 125. It takes four decreases of 25 to get from $n = 100$ to $n = 0$, and $C(100) = 11,000$, so we might estimate $C(0) = 11,000 - 4 \cdot 125 = 10,500$. This means that the fixed cost, before any goods are produced, is $\$10,500$.

9.   (a)

TABLE 2.4

t	0	0.5	1	1.5	2	2.5	3	3.5	4
v(t)	80	64	48	32	16	0	−16	−32	−48

(b)   During the first 2.5 seconds, the velocity is positive, so we know that the rock is headed upward but is slowing down (due to the pull of gravity). After 2.5 seconds, it is falling faster and faster. The negative values of $v(t)$ represent the velocity values of the rock as it is falling downward, back to the ground.

(c) The rock is highest above the ground at the instant before it starts falling downward, which is when the values of the velocity switch from positive to negative. Thus, the velocity will be zero at that instant. Solving

$$v(t) = 0,$$

we obtain

$$80 - 32t = 0,$$

giving

$$t = 2.5 \text{ sec}.$$

We could obtain the same answer by using our table from part (a).

(d) The slope is $\frac{64-80 \text{ ft/sec}}{0.5-0 \text{ sec}} = -32$ ft/sec/sec. Therefore, the acceleration of the rock is -32 ft/sec/sec, since

$$\text{Acceleration} = \frac{\text{Change in velocity}}{\text{Change in time}}.$$

(This is the acceleration due to gravity.)

The t-intercept of 2.5 sec is the instant when the rock goes from going up to falling down, when the velocity is zero. The y-intercept of 80 ft/sec represents the velocity at which the rock was thrown.

(e) Since the pull by the moon's gravitational field is not as strong, the ball would lose speed more and more slowly, so the slope would be less negative. In contrast the pull of the gravitational field of Jupiter is stronger than that of Earth; therefore the slope of $v(t)$ would be a "more negative" number.

13. (a) Since i is linear, we can write

$$i(x) = b + mx.$$

Since $i(10) = 25$ and $i(20) = 50$, we have

$$m = \frac{50 - 25}{20 - 10} = 2.5.$$

So,

$$i(x) = b + 2.5x.$$

Using $i(10) = 25$, we can solve for b:

$$i(10) = b + 2.5(10)$$
$$25 = b + 25$$
$$b = 0.$$

Our formula then is

$$i(x) = 2.5x.$$

(b) The increase in risk associated with *not* smoking is $i(0)$. Since there is no increase in risk for a non-smoker, we have $i(0) = 0$.

(c) The slope of $i(x)$ tells us that the risk increases by a factor of 2.5 with each additional cigarette a person smokes per day.

17. This family of lines all have y–intercept equal to -2. Furthermore, the slopes of these lines are positive. A possible family is shown in Figure 2.11

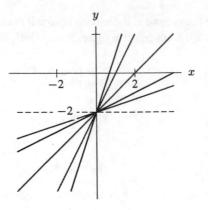

Figure 2.11

21. (a)

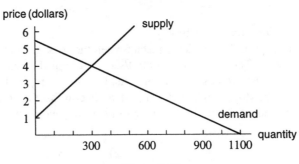

Figure 2.12

(b) The market clearing price occurs where the lines cross. From the graph, it appears as though they cross at $(300, 4)$, which suggests that the market clearing price is $4. This agrees with the answer in Problem 20(d).

25. Answers vary.

CHAPTER THREE

Solutions for Section 3.1

1. (a)

TABLE 3.1

x	-3	-2	-1	0	1	2	3
$f(x)$	1/8	1/4	1/2	1	2	4	8

(b)

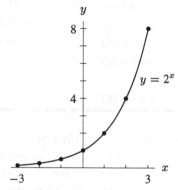

For large negative values of x, $f(x)$ is close to the x-axis. But for large positive values of x, $f(x)$ climbs rapidly away from the x-axis. As x gets larger, y grows more and more rapidly.

5. The population is growing at a rate of 1.9% per year. So, at the end of each year, the population is 100% + 1.9% = 101.9% of what it had been the previous year. The growth factor is 1.019. If P is the population of this country, in millions, and t is the number of years since 1988, then, after one year,

$$P = 70(1.019).$$

After two years, $P = 70(1.019)(1.019) = 70(1.019)^2$

After three years, $P = 70(1.019)(1.019)(1.019) = 70(1.019)^3$

After t years, $P = 70 \underbrace{(1.019)(1.019)\ldots(1.019)}_{t \text{ times}} = 70(1.019)^t$

9. If P represents population and t is the number of years since 1980, then in 1980, $t = 0$ and $P = 222.5$ million. If the population increases by 1.3% per year, then, each year, it is 101.3% of what it had been the year before. So we know that $P = 222.5(1.013)^t$. We want to know t when $P = 350$ million, so we solve $350 = 222.5(1.013)^t$. Using a graph or trial and error calculation, we project that for $t \approx 35.1$ years after 1980, or approximately in the year 2015, the population will have risen to 350 million.

13. Since each filter removes 85% of the remaining impurities, the rate of change of the impurity level is $r = -0.85$ per filter. Thus, the growth factor is $B = 1 + r = 1 - 0.85 = 0.15$. This means that each time the water is passed through a filter, the impurity level L is multiplied by a factor of 0.15. This makes sense, because if each filter removes 85% of the impurities, it will leave behind 15% of the impurities. We see that a formula for L is

$$L = 420(0.15)^n,$$

because after being passed through n filters, the impurity level will have been multiplied by a factor of 0.15 a total of n times.

17. (a) Since for each kilometer above sea level the atmospheric pressure is 86%(100% − 14%) of what it had been one kilometer lower, you could find the pressure at 50 kilometers by taking 86% of 1013 millibars and then taking 86% of your answer and then taking 86% of that result – and repeating the process 50 times. It might make more sense, though, to find a formula. The following table suggests a way to get that formula, with P representing the number of millibars of pressure and h the number of kilometers above sea level.

TABLE 3.2

h	P
0	1013
1	$1013(0.86) = 871.18$
2	$871.18(0.86) = 1013(0.86)(0.86) = 1013(0.86)^2$
3	$1013(0.86)^2 \cdot (0.86) = 1013(0.86)^3$
4	$1013(0.86)^4$
...	...
h	$1013(0.86)^h$

This table rightly suggests that $P = 1013(0.86)^h$. So, at 50 km, $P = 1013(0.86)^{50} \approx 0.538$ millibars.

(b) If we graph the function $P = 1013(0.86)^h$, we can find the value of h for which $P = 900$. One approach is to see where it intersects the line $P = 900$. Doing so, you will see that at an altitude of $P \approx 0.784$ km, the atmospheric pressure will have dropped to 900 millibars.

21. Answers will vary, but they should mention that $f(x)$ is increasing and $g(x)$ is decreasing, that they have the same domain, range, and horizontal asymptote. Some may see that $g(x)$ is a reflection of $f(x)$ about the y-axis whenever $b = \frac{1}{a}$. Graphs might resemble the following:

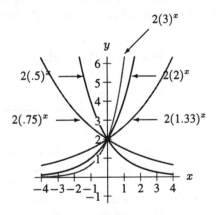

Figure 3.1

Solutions for Section 3.2

1. Let $f(x) = (1.1)^x$, $g(x) = (1.2)^x$, and $h(x) = (1.25)^x$. We note that for $x = 0$,

$$f(x) = g(x) = h(x) = 1;$$

so all three graphs have the same y-intercept. On the other hand, for $x = 1$,

$$f(1) = 1.1, g(1) = 1.2, \quad \text{and} \quad h(1) = 1.25,$$

so $0 < f(1) < g(1) < h(1)$. For $x = 2$,

$$f(2) = 1.21, g(2) = 1.44, \quad \text{and} \quad h(2) = 1.5625,$$

so $0 < f(2) < g(2) < h(2)$. In general, for $x > 0$,

$$0 < f(x) < g(x) < h(x).$$

This suggests that the graph of $f(x)$ lies below the graph of $g(x)$, which in turn lies below the graph of $h(x)$, and that all lie above the x-axis. Alternately, you can consider 1.1, 1.2, and 1.25 as growth factors to conclude $y = (1.25)^x$ is the top function, and $y = (1.2)^x$ is in the middle.

5. (a) If a function is linear, then the differences in successive function values will be constant. If a function is exponential, the ratios of successive function values will remain constant. Now

$$f(1) - f(0) = 13.75 - 12.5 = 1.25$$

while

$$f(2) - f(1) = 15.125 - 13.75 = 1.375.$$

Thus, $f(x)$ is not linear. On the other hand,

$$\frac{f(1)}{f(0)} = \frac{13.75}{12.5} = 1.1$$

and

$$\frac{f(2)}{f(1)} = \frac{15.25}{13.75} = 1.1.$$

Checking the rest of the data, we see that the ratios of differences remains constant, so $f(x)$ is exponential.

Now

$$g(1) - g(0) = 2 - 0 = 2$$

and

$$g(2) - g(1) = 4 - 2 = 2.$$

Checking the rest of the data, we see that the differences remain constant, so $g(x)$ is linear.

Now

$$h(1) - h(0) = 12.6 - 14 = -1.4$$

while

$$h(2) - h(1) = 11.34 - 12.6 = -1.26.$$

Thus, $h(x)$ is not linear. On the other hand,

$$\frac{h(1)}{h(0)} = 0.9$$

$$\frac{h(2)}{h(1)} = \frac{11.34}{12.6} = 0.9.$$

Checking the rest of the data, we see that the ratio of differences remains constant, so $h(x)$ is exponential.

Now

$$i(1) - i(0) = 14 - 18 = -4$$

and

$$i(2) - i(1) = 10 - 14 = -4.$$

Checking the rest of the data, we see that the differences remain constant, so $i(x)$ is linear.

(b) We know that f is exponential, so

$$f(x) = AB^x$$

for some constants A and B. We know that $f(0) = 12.5$, so

$$12.5 = f(0)$$
$$12.5 = AB^0$$
$$12.5 = A(1).$$

Thus,

$$A = 12.5.$$

We also know

$$13.75 = f(1)$$
$$13.75 = 12.5B.$$

Thus,

$$B = \frac{13.75}{12.5} = 1.1.$$

As a result,

$$f(x) = 12.5(1.1)^x.$$

The graph of $f(x)$ looks like

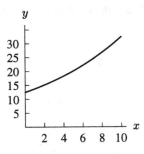

Figure 3.2

We know that $g(x)$ is linear, so it must be of the form

$$g(x) = b + mx$$

where m is the slope and b is the y-intercept. Since at $x = 0$, $g(0) = 0$, we know that the y-intercept is 0, so $b = 0$. Using the points $(0, 0)$ and $(1, 2)$, the slope is

$$m = \frac{2 - 0}{1 - 0} = 2.$$

Thus,

$$g(x) = 0 + 2x = 2x.$$

The graph of $y = g(x)$ looks like

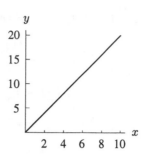

Figure 3.3

9. One approach to the problem is to graph both functions and to see where the graph of $p(x)$ is below the graph of $q(x)$.

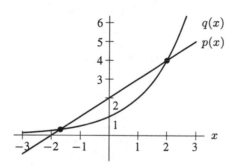

From the graph, we see that $p(x)$ intersects $q(x)$ in two places; namely, at $x \approx -1.7$ and $x = 2$. We notice that $p(x)$ is above $q(x)$ between these two points and below $q(x)$ outside the segment defined by these two points. Hence $p(x) < q(x)$ for $x < -1.7$ and for $x > 2$.

13. (a) Since $h(x) = AB^x$, $h(0) = AB^0 = A(1) = A$. We are given $h(0) = 3$, so $A = 3$. If $h(x) = 3B^x$, then $h(1) = 3B^1 = 3B$. But we are told that $h(1) = 15$, so $3B = 15$ and $B = 5$. Therefore $h(x) = 3(5)^x$.

    (b) Since $f(x) = AB^x$, $f(3) = AB^3$ and $f(-2) = AB^{-2}$. Since we know that $f(3) = -\frac{3}{8}$ and $f(-2) = -12$, we can say

$$AB^3 = -\frac{3}{8}$$

and

$$AB^{-2} = -12.$$

Forming ratios, we have

$$\frac{AB^3}{AB^{-2}} = \frac{-\frac{3}{8}}{-12}$$

$$B^5 = -\frac{3}{8} \times -\frac{1}{12} = \frac{1}{32}.$$

Since $32 = 2^5$, $\frac{1}{32} = \frac{1}{2^5} = (\frac{1}{2})^5$. This tells us that

$$B = \frac{1}{2}.$$

Thus, our formula is $f(x) = A(\frac{1}{2})^x$. Use $f(3) = A(\frac{1}{2})^3$ and $f(3) = -\frac{3}{8}$ to get

$$A(\frac{1}{2})^3 = -\frac{3}{8}$$
$$A(\frac{1}{8}) = -\frac{3}{8}$$
$$\frac{A}{8} = -\frac{3}{8}$$
$$A = -3.$$

Therefore $f(x) = -3(\frac{1}{2})^x$.

(c) Since $g(x) = AB^x$, we can say that $g(\frac{1}{2}) = AB^{\frac{1}{2}}$ and $g(\frac{1}{4}) = AB^{\frac{1}{4}}$. Since we know that $g(\frac{1}{2}) = 4$ and $g(\frac{1}{4}) = 2\sqrt{2}$, we can conclude that

$$AB^{\frac{1}{2}} = 4 = 2^2$$

and

$$AB^{\frac{1}{4}} = 2\sqrt{2} = 2 \cdot 2^{\frac{1}{2}} = 2^{\frac{3}{2}}.$$

Forming ratios, we have

$$\frac{AB^{\frac{1}{2}}}{AB^{\frac{1}{4}}} = \frac{2^2}{2^{\frac{3}{2}}}$$
$$B^{\frac{1}{4}} = 2^{\frac{1}{2}}$$
$$(B^{\frac{1}{4}})^4 = (2^{\frac{1}{2}})^4$$
$$B = 2^2 = 4.$$

Now we know that $g(x) = A(4)^x$, so $g(\frac{1}{2}) = A(4)^{\frac{1}{2}} = 2A$. Since we also know that $g(\frac{1}{2}) = 4$, we can say

$$2A = 4$$
$$A = 2.$$

Therefore $g(x) = 2(4)^x$.

17. Since the function is exponential, we know that $y = AB^x$. Since $(0, 1.2)$ is on the graph, we know $1.2 = AB^0$, and that $A = 1.2$. To find B, we use point $(2, 4.8)$ which gives

$$4.8 = 1.2(B)^2$$
$$4 = B^2$$
$$B = 2, \text{ since } B > 0.$$

Thus, $y = 1.2(2)^x$ is a possible formula for this function.

21. (a)

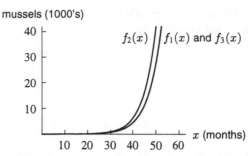

mussels (1000's)

From this graph, the three models seem to be in good agreement. Models 1 and 3 are indistinguishable; model 2 appears to rise a little faster. However notice that we cannot see the behavior beyond 50 months because our function values go beyond the top of the viewing window.

(b)

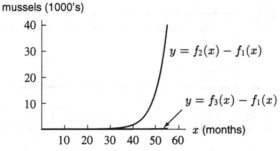

mussels (1000's)

The graph of $y = f_2(x) - f_1(x) = 3(1.21)^x - 3(1.2)^x$ grows very rapidly, especially after 40 months. The graph of $y = f_3(x) - f_1(x) = 3.01(1.2)^x - 3(1.2)^x$ is hardly visible on this scale.

(c) Models 1 and 3 are in good agreement, but model 2 predicts a much larger mussel population than does model 1 after only 50 months. We can come to at least two conclusions. First, even small differences in the base of an exponential function can be highly significant, while differences in initial values are not as significant. Second, although two exponential curves can look very similar, they can actually be making very different predictions as time increases.

25. (a) Since the population is growing by a certain percent each year, we know that it can be described by the formula $P = AB^t$. If t is the number of years since 1953, then A will represent the population in 1953. To find A, we will substitute the values we know into the formula. If the growth rate is 8%, then each year the population is multiplied by the growth factor 1.08, so $B = 1.08$. Thus,

$$P = A(1.08)^t.$$

We know that in 1993 ($t = 40$) the population was 13 million, so

$$13,000,000 = A(1.08)^{40}$$
$$13,000,000 = A(21.72)$$
$$A = \frac{13,000,000}{21.72} \approx 600,000.$$

Therefore in 1953, the population of humans in Florida was about 600,000 people.

(b) In 1953 ($t = 0$), the bear population was 11,000, so $A = 11,000$. The population has been decreasing at a rate of 6% a year, so the growth rate is $100\% - 6\% = 94\%$ or 0.94. Thus, the growth function for black bears is

$$P = (11,000)(0.94)^t.$$

In 1993, $t = 40$, so

$$P = (11,000)(0.94)^{40} \approx 926.$$

(c) To find the year t when the bear population will be 100, we set P equal to 100 in the equation found in part (b) and get an equation involving t:

$$P = (11,000)(0.94)^t$$
$$100 = (11000)(0.94)^t$$
$$\frac{100}{11000} = (0.94)^t$$
$$0.00909 \approx (0.94)^t.$$

By looking at the intersection of the graphs $P = 0.00909$ and $P = (0.94)^t$, or by trial and error, we find that $t \approx 76$ years. Our model predicts that in 76 years from 1953, which is the year 2029, the population of black bears will fall below 100.

Solutions for Section 3.3

1.

TABLE 3.3

n	1	2	3	4	5	6	7	8	9
$\log n$	0	0.3010	0.4771	0.6021	0.6990	0.7782	0.8451	0.9031	0.9542

TABLE 3.4

n	10	20	30	40	50	60	70	80	90
$\log n$	1	1.3010	1.4771	1.6021	1.6990	1.7782	1.8451	1.9031	1.9542

TABLE 3.5

n	100	200	300	400	500	600	700	800	900
$\log n$	2	2.3010	2.4771	2.6021	2.6990	2.7782	2.8451	2.9031	2.9542

5. (a) Figure 3.4 shows the track events plotted on a linear scale.

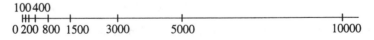

Figure 3.4:

(b) Figure 3.5 shows the track events plotted on a logarithmic scale.

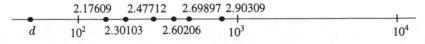

Figure 3.5:

(c) Figure 3.4 gives a runner better information about pacing for the distance.
(d) On Figure 3.4 the point 50 is $\frac{1}{2}$ the distance from 0 to 100. On Figure 3.5 the point 50 is the same distance to the left of 100 as 200 is to the right. This is shown as point d.

Solutions for Section 3.4

1. As we do these problems, keep in mind that we are looking for a power of 10. For example, log 10,000 is asking for the power of 10 which will give 10,000. Since $10^4 = 10,000$, we know that log 10,000 = 4.

   (a) Since $1 = 10^0$, log 1 = 0.

   (b) Since $0.1 = \frac{1}{10} = 10^{-1}$, we know that $\log 0.1 = \log 10^{-1} = -1$.

   (c) In this problem, we can use the identity $\log 10^N = N$. So log $10^0 = 0$. We can check this by observing that $10^0 = 1$, similar to what we saw in (a), that log 1 = 0.

   (d) To find the $\log \sqrt{10}$ we need to recall that $\sqrt{10} = 10^{1/2}$. Now we can use our identity and say
   $$\log \sqrt{10} = \log 10^{1/2} = \frac{1}{2}.$$

   (e) Using the identity, we get $\log 10^5 = 5$.

   (f) Using the identity, we get $\log 10^2 = 2$.

   (g) $\log \dfrac{1}{\sqrt{10}} = \log 10^{-1/2} = -\dfrac{1}{2}$

   For the last three problems, we'll use the identity $10^{\log N} = N$.

   (h) $10^{\log 100} = 100$

   (i) $10^{\log 1} = 1$

   (j) $10^{\log 0.01} = 0.01$

5. (a) True.

   (b) False. $\log A \log B = \log A \cdot \log B$, not $\log A + \log B$.

   (c) False. $\frac{\log A}{\log B}$ cannot be rewritten.

   (d) True.

   (e) True. $\sqrt{x} = x^{1/2}$ and $\log x^{1/2} = \frac{1}{2} \log x$.

   (f) False. $\sqrt{\log x} = (\log x)^{1/2}$.

9. (a)
$$\log 100^x = \log(10^2)^x$$
$$= \log 10^{2x}.$$

Since $\log 10^N = N$, then
$$\log 10^{2x} = 2x.$$

   (b)
$$1000^{\log x} = (10^3)^{\log x}$$
$$= (10^{\log x})^3$$

Since $10^{\log x} = x$ we know that
$$(10^{\log x})^3 = (x)^3 = x^3.$$

   (c)
$$\log 0.001^x = \log \left(\frac{1}{1000}\right)^x$$
$$= \log(10^{-3})^x$$
$$= \log 10^{-3x}$$
$$= -3x.$$

13. (a) $\log(\log 10) = \log 1 = 0.$

(b) Substituting 10^2 for 100 we have

$$\sqrt{\log 100} - \log \sqrt{100} = \sqrt{\log 10^2} - \log \sqrt{10^2}$$

Since $\log 10^2 = 2$ and $\sqrt{10^2} = 10$ we have

$$\sqrt{\log 10^2} - \log \sqrt{10^2} = \sqrt{2} - \log 10$$

But $\log 10 = 1$, so

$$\sqrt{2} - \log 10 = \sqrt{2} - 1.$$

(c) We will first simplify the expression $\sqrt{10}\sqrt[3]{10}\sqrt[5]{10}$ by using exponents instead of radicals:

$$
\begin{aligned}
\sqrt{10}\sqrt[3]{10}\sqrt[5]{10} &= 10^{\frac{1}{2}} \cdot 10^{\frac{1}{3}} \cdot 10^{\frac{1}{5}} \\
&= 10^{\frac{1}{2}+\frac{1}{3}+\frac{1}{5}} \quad \text{(using an exponent rule)} \\
&= 10^{\frac{15+10+6}{30}} \quad \text{(finding an LCD)} \\
&= 10^{31/30}.
\end{aligned}
$$

Thus,

$$\log \sqrt{10}\sqrt[3]{10}\sqrt[5]{10} = \log 10^{31/30} = \frac{31}{30}.$$

(d)

$$
\begin{aligned}
1000^{\log 3} &= (10^3)^{\log 3} \\
&= (10^{\log 3})^3 \quad \text{(using an exponent rule)} \\
&= 3^3 \quad \text{(definition of log 3)} \\
&= 27.
\end{aligned}
$$

(e)

$$
\begin{aligned}
0.01^{\log 2} &= \left(\frac{1}{100}\right)^{\log 2} \\
&= (10^{-2})^{\log 2} \\
&= (10^{\log 2})^{-2} \\
&= 2^{-2} = \frac{1}{4}.
\end{aligned}
$$

(f)

$$
\begin{aligned}
\frac{1}{\log \frac{1}{\log \sqrt[10]{10}}} &= \frac{1}{\log \frac{1}{\log 10^{1/10}}} \\
&= \frac{1}{\log \frac{1}{(1/10)}} \quad \text{(because } \log 10^{1/10} = \tfrac{1}{10}) \\
&= \frac{1}{\log 10} \quad \text{(since } \tfrac{1}{1/10} = 10) \\
&= 1 \quad \text{(because } \log 10 = 1)
\end{aligned}
$$

17. Let $P = AB^t$ where P is the number of bacteria at time t hours since the beginning of the experiment. A is the number of bacteria we're starting with.

(a) Since the colony begins with 3 bacteria we have $A = 3$. Using the information that $P = 100$ when $t = 3$, we can solve the following equation for B:

$$P = 3B^t$$
$$100 = 3B^3$$
$$\sqrt[3]{\frac{100}{3}} = B$$
$$B = \left(\frac{100}{3}\right)^{1/3} \approx 3.22$$

Therefore, $P = 3(3.22)^t$.

(b) We want to find the value of t for which the population triples, going from three bacteria to nine. So we want to solve:

$$9 = 3\left(\frac{100}{3}\right)^{\frac{t}{3}}$$
$$3 = \left(\frac{100}{3}\right)^{\frac{t}{3}}$$
$$3^3 = \left(\left(\frac{100}{3}\right)^{\frac{t}{3}}\right)^3$$
$$27 = \left(\frac{100}{3}\right)^t$$
$$\log 27 = \log\left(\frac{100}{3}\right)^t$$
$$\log 27 = t\log\left(\frac{100}{3}\right)$$
$$\frac{\log 27}{\log\left(\frac{100}{3}\right)} = t$$

$$t = 0.9399 \approx 0.94 \text{ hours, or about 56.4 minutes.}$$

21. If t represents the number of years since 1990, let $W(t)$ = population of Erehwon at time t, in millions of people, and let $C(t)$ = population of Ecalpon at time t, in millions of people. Since the population of both Erehwon and Ecalpon are increasing by a constant percent, we know that they are both exponential functions. In Erehwon, the growth factor is 1.029. Since its population in 1990 (when $t = 0$) is 50 million people, we know that $W(t) = 50(1.029)^t$ (see Table 3.6 to see this formula developed). In Ecalpon, the growth factor is 1.032, and starts at 45 million, so $C(t) = 45(1.032)^t$.

TABLE 3.6

t	$W(t)$
0	50
1	102.9% of $50 = 50(1.029)$
2	102.9% of $50(1.029) = 50(1.029)(1.029) = 50(1.029)^2$
3	102.9% of $50(1.029)^2 = 50(1.029)^3$
4	$50(1.029)^4$
⋮	⋮
t	$50(1.029)^t$

(a) The formula for Erehwon and Ecalpon are, respectively:

$$W(t) = 50(1.029)^t$$
$$C(t) = 45(1.032)^t$$

(b) The two countries will have the same population when $W(t) = C(t)$. We therefore need to solve:

$$50(1.029)^t = 45(1.032)^t$$

$$\frac{1.032^t}{1.029^t} = \left(\frac{1.032}{1.029}\right)^t = \frac{50}{45} = \frac{10}{9}$$

$$\log\left(\frac{1.032}{1.029}\right)^t = \log\left(\frac{10}{9}\right)$$

$$t\log\left(\frac{1.032}{1.029}\right) = \log\left(\frac{10}{9}\right)$$

$$t = \frac{\log\left(\dfrac{10}{9}\right)}{\log\left(\dfrac{1.032}{1.029}\right)} \approx 36.2$$

So the populations are equal after about 36.2 years, in the year 2026.

(c) The population of Ecalpon is double the population of Erehwon when

$$C(t) = 2W(t)$$

that is, when

$$45(1.032)^t = 2 \cdot 50(1.029)^t.$$

We will use logs to help us solve the equation.

$$45(1.032)^t = 100(1.029)^t$$

$$\frac{(1.032)^t}{(1.029)^t} = \frac{100}{45} = \frac{20}{9}$$

$$\left(\frac{1.032}{1.029}\right)^t = \frac{20}{9}$$

$$\log\left(\frac{1.032}{1.029}\right)^t = \log\left(\frac{20}{9}\right)$$

$$t \log \left(\frac{1.032}{1.029}\right) = \log \left(\frac{20}{9}\right)$$

$$t = \frac{\log \left(\frac{20}{9}\right)}{\log \left(\frac{1.032}{1.029}\right)} \approx 274 \text{ years.}$$

So it will take about 274 years for the population of Ecalpon to be twice that of Erehwon.

Solutions for Section 3.5

1. (a) Let the functions graphed in (a), (b), and (c) be called $f(x)$, $g(x)$, and $h(x)$ respectively. Looking at the graph of $f(x)$, we see that $f(10) = 3$. In the table for $r(x)$ we note that $r(10) = 1.6990$ so $f(x) \neq r(x)$. Similarly, $s(10) = 0.6990$, so $f(x) \neq s(x)$. The values describing $t(x)$ do seem to satisfy the graph of $f(x)$, however. In the graph, we note that when $0 < x < 1$, then y must be negative. The data point $(0.1, -3)$ satisfies this. When $1 < x < 10$, then $0 < y < 3$. In the table for $t(x)$, we see that the point $(2, 0.9031)$ satisfies this condition. Finally, when $x > 10$ we see that $y > 3$. The values $(100, 6)$ satisfy this. Therefore, $f(x)$ and $t(x)$ could represent the same function.

   (b) For $g(x)$, we note that
   $$\begin{cases} \text{when } 0 < x < 0.2, & \text{then } y < 0; \\ \text{when } 0.2 < x < 1, & \text{then } 0 < y < 0.699; \\ \text{when } x > 1, & \text{then } y > 0.699. \end{cases}$$
   All the values of x in the table for $r(x)$ are greater than 1 and all the corresponding values of y are greater than 0.699, so $g(x)$ could equal $r(x)$. We see that, in $s(x)$, the values $(0.5, -0.06021)$ do not satisfy the second condition so $g(x) \neq s(x)$. Since we already know that $t(x)$ corresponds to $f(x)$, we conclude that $g(x)$ and $r(x)$ correspond.

   (c) By elimination, $h(x)$ must correspond to $s(x)$. We see that in $h(x)$,
   $$\begin{cases} \text{when } x < 2, & \text{then } y < 0; \\ \text{when } 2 < x < 20, & \text{then } 0 < y < 1; \\ \text{when } x > 20, & \text{then } y > 1. \end{cases}$$
   Since the values in $s(x)$ satisfy these conditions, it is reasonable to say that $h(x)$ and $s(x)$ correspond.

5. (a) Use $o(t)$ to describe the number of owls as a function of time. After 1 year, we see that the number of owls is 103% of 245, or $o(1) = 245(1.03)$. After 2 years, the population is 103% of that number, or $o(2) = (245(1.03)) \cdot 1.03 = 245(1.03)^2$. After t years, it is $o(t) = 245(1.03)^t$.

   (b) We will use $h(t)$ to describe the number of hawks as a function of time. Since $h(t)$ doubles every 10 years, we know that its growth factor is constant and so it is an exponential function with a formula of the form $h(t) = AB^t$. In this case the initial population is 63 hawks, so $h(t) = 63B^t$. We are told that the population in 10 years, $h(t + 10)$, is twice the current population, that is

   $$h(t + 10) = 2h(t).$$

   But $h(t) = 63B^t$ and $h(t + 10) = 63B^{(t+10)}$, so
   $$63 \cdot B^{t+10} = 2 \cdot (63 \cdot B^t) = 2 \cdot 63 \cdot B^t$$
   $$B^{t+10} = 2 \cdot B^t$$
   $$\frac{B^{t+10}}{B^t} = 2$$
   $$B^{10} = 2$$
   $$B = 2^{1/10} \approx 1.07.$$

   Thus, the number of hawks as a function of time is
   $$h(t) = 63 \cdot (1.07)^t.$$

(c)   Looking at Figure 3.6 we see that it takes about 34.2 years for the populations to be equal.

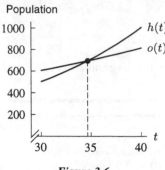

Figure 3.6

Solutions for Section 3.6

1. (a)   The nominal interest rate is 8%, so the interest rate per month is .08/12. Therefore, at the end of 3 years, or 36 months,

$$\text{Balance} = \$1000 \left(1 + \frac{0.08}{12}\right)^{36} = \$1270.24.$$

(b)   There are 52 weeks in a year, so the interest rate per week is .08/52. At the end of $52 \times 3 = 156$ weeks,

$$\text{Balance} = \$1000 \left(1 + \frac{0.08}{52}\right)^{156} = \$1271.01.$$

(c)   Assuming no leap years, the interest rate per day is .08/365. At the end of 3×365 days

$$\text{Balance} = \$1000 \left(1 + \frac{0.08}{365}\right)^{3 \cdot 365} = \$1271.22.$$

(d)   With continuous compounding, after 3 years

$$\text{Balance} = \$1000 e^{0.08(3)} = \$1271.25$$

5.   Since the first student's $500 is growing by a factor of 1.045 each year $(100\% + 4.5\%)$, a formula that describes how much money she has at the end of t years is $A_1 = 500(1.045)^t$. A formula for the second student's investment is $A_2 = 800(1.03)^t$. We need to find the value of t for which $A_1 = A_2$. That is, when

$$500(1.045)^t = 800(1.03)^t$$

$$\frac{1.045^t}{1.03^t} = \frac{800}{500}$$

$$\left(\frac{1.045}{1.03}\right)^t = \frac{8}{5}$$

$$\log\left(\frac{1.045}{1.03}\right)^t = \log\left(\frac{8}{5}\right)$$

$$t \log\left(\frac{1.045}{1.03}\right) = \log\left(\frac{8}{5}\right)$$

$$t = \frac{\log\left(\frac{8}{5}\right)}{\log\left(\frac{1.045}{1.03}\right)} \approx 32.5.$$

The balances will be equal in about 32.5 years.

9. (a) The effective annual yield is the rate at which the account is actually increasing in one year. According to the formula, $M = M_0(1.07763)^t$, at the end of one year you have $M = 1.07763M_0$, or 1.07763 times what you had the previous year. The account is 107.763% larger than it had been previously; that is, it increased by 7.763%. Thus the effective yield being paid on this account each year is about 7.76%.

    (b) Since the money is being compounded each month, one way to find the nominal annual rate is to determine the rate being paid each month. In t years there are $12t$ months, and so, if b is the monthly growth factor, our formula becomes

$$M = M_0 b^{12t} = M_0(b^{12})^t.$$

Thus, equating the two expressions for M, we see that

$$M_0(b^{12})^t = M_0(1.07763)^t.$$

Dividing both sides by M_0 yields
$$(b^{12})^t = (1.07763)^t.$$

Taking the tth root of both sides, we have

$$b^{12} = 1.07763$$

which means that
$$b = (1.07763)^{1/12} \approx 1.00625.$$

Thus, this account earns 0.625% interest every month, which amounts to a nominal interest rate of about $12(0.625\%) = 7.5\%$.

13. If an investment decreases by 5% each year, we know that only 95% remains at the end of the first year. After 2 years there will be 95% of 95%, or 0.95^2 left. After 4 years, there will be $0.95^4 \approx 0.8145$ or 81.45% of the investment left; it therefore decreases by about 18.55% altogether.

17. (a) Since $P(t)$ has continuous growth, its formula will be $P(t) = P_0 e^{kt}$. Since P_0 is the initial population, which is 22,000, and k represents the continuous growth rate of 7.1%, our formula is

$$P(t) = 22,000e^{0.071t}.$$

    (b) While, at any given instant, the population is growing at a rate of 7.1% a year, the effect of compounding is to give us an actual increase of more than 7.1%. To find that increase, we first need to find the growth factor, or B. Rewriting $P(t) = 22,000e^{0.071t}$ in the form $P = 22000B^t$ will help us accomplish this. Thus, $P(t) = 22,000(e^{0.071})^t \approx 22,000(1.0736)^t$. Alternatively, we can equate the two formulas and solve for B:

$$22,000e^{0.071t} = 22,000B^t$$
$$e^{0.071t} = B^t \quad \text{(dividing both sides by 22,000)}$$
$$e^{0.071} = B \quad \text{(taking the } t^{\text{th}} \text{ root of both sides).}$$

Using your calculator, you can find that $B \approx 1.0736$. Either way, we see that at the end of the year, the population is 107.36% of what it had been at the end of the previous year, and so the population increases by approximately 7.36% each year.

21. For investment A, $P = 875(1 + \frac{0.135}{365})^{365(2)} = \$1146.16$. For investment B, $P = 1000(e^{0.067(2)}) = \$1143.39$. For investment C, $P = 1050(1 + \frac{0.045}{12})^{12(2)} = \$1148.69$ So from best to worse we have C, A, and B.

25. To find the fee for six hours, we need to find the hourly rate of interest. If it is 20% per year, then it is
    (20%/year)·(1 year/365 days)·(1 day/24 hours)= $\frac{20\%}{(365)(24)}$ per hour.

    Since the interest is being compounded continuously, the total amount of money is described by $P = P_0 e^{kt}$, where, in this case, k is the hourly rate and t is the number of hours. So

    $$P = 200,000,000 e^{\frac{0.20}{(365)(24)}(6)} = 200,027,399.1$$

    The value of the money at the end of the six hours was \$200,027,399.10, so the fee for that time was \$27,399.10.

Solutions for Section 3.7

1. (a) The number of bacteria present after 1/2 hour is

    $$N = 1000 e^{0.69(1/2)} \approx 1412.$$

    If you notice that $0.69 \approx \ln 2$, you could also say

    $$N = 1000 e^{0.69/2} \approx 1000 e^{\frac{1}{2} \ln 2} = 1000 e^{\ln \sqrt{2}} = 1000\sqrt{2} \approx 1412.$$

    (b) We solve for t in the equation

    $$1,000,000 = 1000 e^{0.69t}$$
    $$e^{0.69t} = 1000$$
    $$0.69t = \ln 1000$$
    $$t = \left(\frac{\ln 1000}{0.69}\right) \approx 10.0 \text{ hours.}$$

    (c) The doubling time is the time T such that $N = 2000$, so

    $$2000 = 1000 e^{0.69T}$$
    $$e^{0.69T} = 2$$
    $$0.69T = \ln 2$$
    $$T = \left(\frac{\ln 2}{0.69}\right) \approx 1.0 \text{ hours.}$$

    If you notice that $0.69 \approx \ln 2$, you see why the half-life turns out to be 1 hour:

    $$e^{0.69T} = 2$$
    $$e^{T \ln 2} \approx 2$$
    $$e^{\ln 2^T} \approx 2$$
    $$2^T \approx 2$$
    $$T \approx 1$$

5. Since the formula for finding the value, $P(t)$, of an \$800 investment after t years at 4% interest compounded annually is $P(t) = 800(1.04)^t$ and we want to find the value of t when $P(t) = 2,000$, we must solve:

$$800(1.04)^t = 2000$$
$$1.04^t = \frac{2000}{800} = \frac{20}{8} = \frac{5}{2}$$
$$\log 1.04^t = \log \frac{5}{2}$$
$$t \log 1.04 = \log \frac{5}{2}$$
$$t = \frac{\log(5/2)}{\log 1.04} \approx 23.4 \text{ years.}$$

So it will take about 23.4 years for the \$800 to grow to \$2,000.

9. (a) For a function of the form Ae^{rt}, A is the population at $t = 0$ and r is the rate of growth. So the growth rate is 3%.

(b) In year $t = 0$, the population is $N(0) = 5.3$ million.

(c) We want to find t such that the population of 5.3 million triples to 15.9 million. So, for what value of t does $N(t) = 5.3e^{0.03t} = 15.9$?

$$5.3e^{0.03t} = 15.9$$
$$e^{0.03t} = 3$$
$$\ln e^{0.03t} = \ln 3$$
$$0.03t = \ln 3$$
$$t = \frac{\ln 3}{0.03} \approx 36.6$$

So the population will triple in approximately 36.6 years.

(d) Since $N(t)$ is in millions, we want to find t such that $N(t) = 0.000001$.

$$5.3e^{0.03t} = 0.000001$$
$$e^{0.03t} = \frac{0.000001}{5.3} \approx 0.000000189$$
$$\ln e^{0.03t} \approx \ln(0.000000189)$$
$$0.03t \approx \ln(0.000000189)$$
$$t \approx \frac{\ln(0.000000189)}{0.03} \approx -516$$

According to this model, the population of Washington State was 1 person 516 years ago. It is unreasonable to suppose the formula extends so accurately into the past. It is also unlikely that exactly one person was ever in Washington State.

13. (a) The population has increased by $34,000 - 30,000 = 4,000$ people in that time period, so its total percent increase is $\frac{4,000}{30,000} = 0.133 = 13.3\%$.

(b) If B represents the annual growth factor, then in five years the population will have grown by a factor of B^5. We learned in part (a) that the population has increased by 13.3% in that time, so it is 113.3% of what it had been five years earlier. Thus

$$B^5 = 1.133$$
$$B = (1.133)^{\frac{1}{5}} \approx 1.0253.$$

If, at the end of each year, the population is 102.53% of what it had been at the beginning of the year, then the rate of growth is about 2.53% per year.

(c)   The continuous annual growth rate is represented by k in the formula $P(t) = P_0 e^{kt}$. Since we know that the initial population is 30,000 and the growth factor is 102.53 (from (b)), we can say that $P(t) = 30,000(1.0253)^t$ defines this function. To find k, we can equate the two formulas:

$$30,000(1.0253)^t = P_0 e^{kt} = 30,000e^{kt}$$
$$1.0253^t = e^{kt}$$
$$1.0253 = e^k$$
$$\ln 1.0253 = \ln e^k$$
$$\ln 1.0253 = k$$
$$k \approx 0.0250.$$

Thus, while the population is growing at 2.53% per year, it is growing at a rate of 2.50% at any given instant.

17. (a)   A graph of this equation is shown in Figure 3.7.

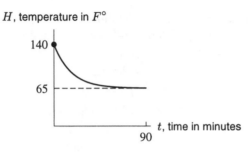

Figure 3.7

(b)   The H-intercept is $(0, 140)$. This is the temperature of the coffee at the time it is poured. This agrees with the results from our formula if we evaluate H at $t = 0$:

$$H = 75e^{-0.06(0)} + 65 = 75e^0 + 65 = 75(1) + 65 = 140.$$

(c)   At 8:10 AM, ten minutes after I poured it, the temperature of the coffee is $H(10) = 75e^{-0.06(10)} + 65 \approx$ 106°. At 9 AM, an hour after being poured, the temperature is $H(60) = 75e^{-0.06(60)} + 65 \approx 67°$.

(d)   Finding values of $H(t)$ as t gets larger and larger, we see that $H(t)$ is getting closer and closer to 65°. On the graph, this corresponds to the horizontal asymptote at $H = 65$.

(e)   In order to find the outer limits of pleasurable coffee drinking, we need to solve the following two equations:

$$125 = 75e^{-0.06t} + 65,$$

and

$$100 = 75e^{-0.06t} + 65.$$

These can be solved for t but since we already have a graph, let's find the points of intersection of

$$H = 125 \text{ and } H = 75e^{-0.06t} + 65$$
$$H = 100 \text{ and } H = 75e^{-0.06t} + 65.$$

In the first case, it is $(3.7, 125)$, while in the second it is $(12.7, 100)$. So we know that it will take about 3.7 minutes for the temperature to get down to 125°F, while it will hit 100°F in 12.7 minutes after it is poured. So, we can begin to drink our coffee 3.7 minutes after it is poured and we have $12.7 - 3.7 = 9$ minutes to drink it.

21. (a)

$$e^{x+4} = 10$$
$$\ln e^{x+4} = \ln 10$$
$$x + 4 = \ln 10$$
$$x = \ln 10 - 4$$

(b)

$$e^{x+5} = 7 \cdot 2^x$$
$$\ln e^{x+5} = \ln(7 \cdot 2^x)$$
$$x + 5 = \ln 7 + \ln 2^x$$
$$x + 5 = \ln 7 + x \ln 2$$
$$x - x \ln 2 = \ln 7 - 5$$
$$x(1 - \ln 2) = \ln 7 - 5$$
$$x = \frac{\ln 7 - 5}{1 - \ln 2}$$

(c) $\log(2x + 5) \cdot \log(9x^2) = 0$

In order for this product to equal zero, we know that one or both terms must be equal to zero. Thus, we will set each of the factors equal to zero to determine the values of x for which the factors will equal zero. We have

$$\log(2x + 5) = 0 \qquad \text{or} \qquad \log(9x^2) = 0$$
$$2x + 5 = 1 \qquad\qquad\qquad 9x^2 = 1$$
$$2x = -4 \qquad\qquad\qquad x^2 = \frac{1}{9}$$
$$x = -2 \qquad\qquad\qquad x = \frac{1}{3} \text{ or } x = -\frac{1}{3}.$$

Thus our solutions are $x = -2$, $\frac{1}{3}$, or $-\frac{1}{3}$.

25. (a) Let t be the number of years in a man's age above 30 (i.e. let $t =$ the man's age-30) and let M_0 denote his bone mass at age 30. If he is losing 2% per year, then 98% remains after each year, and thus we can say that $M(t) = M_0(0.98)^t$, where $M(t)$ represents the man's bone mass t years after age 30. But we want a formula describing bone mass in terms of a, his age. Since t is number of years in his age over 30, $t = a - 30$. So, we can substitute $a - 30$ for t in our formula to find an expression in terms of a:

$$M(a) = M_0(0.98)^{(a-30)}.$$

(b) We want to know for what value of a

$$M(a) = \frac{1}{2}M_0$$

Therefore, we will solve $\quad M_0(0.98)^{(a-30)} = \frac{1}{2}M_0$

$$(0.98)^{(a-30)} = \frac{1}{2}$$
$$\log\left((0.98)^{(a-30)}\right) = \log\frac{1}{2} = \log 0.5$$

$$(a - 30)\log(0.98) = \log 0.5$$

$$a - 30 = \frac{\log 0.5}{\log 0.98}$$

$$a = 30 + \frac{\log 0.5}{\log(0.98)} \approx 64.3$$

The average man will have lost half his bone mass at approximately 64.3 years of age.

29. (a)

$$\ln(x) - \ln(100 - x) = 0.48t - \ln(99)$$
$$\ln(x) - \ln(100 - x) + \ln 99 = 0.48t$$
$$0.48t = \ln x - \ln(100 - x) + \ln 99$$
$$0.48t = ((\ln(x) + \ln 99) - \ln(100 - x))$$

Using the laws of logarithms, i.e $\log a + \log b = \log(ab)$ and $\log a - \log b = \log\left(\frac{a}{b}\right)$, we can rewrite the right side of the equation as follows:

$$0.48t = \ln\left(\frac{99x}{100 - x}\right)$$

$$t = \frac{1}{0.48}\ln\left(\frac{99x}{100 - x}\right)$$

(b) From the given equation we have

$$\ln(x) - \ln(100 - x) = 0.48t - \ln 99$$

Using the laws of logarithms, we have

$$\ln\left(\frac{x}{100 - x}\right) = 0.48t - \ln 99$$

Raising both sides to the e power we have

$$e^{\ln\left(\frac{x}{100-x}\right)} = e^{0.48t - \ln 99}$$

Since $e^{\ln x} = x$, we have

$$\frac{x}{100 - x} = e^{0.48t - \ln 99}$$

$$x = (100 - x)(e^{0.48t - \ln 99})$$

$$x = 100e^{0.48t - \ln 99} - xe^{0.48t - \ln 99}$$

$$x + xe^{0.48t - \ln 99} = 100e^{0.48t - \ln 99}$$

$$x(1 + e^{0.48t - \ln 99}) = 100e^{0.48t - \ln 99}$$

$$x = \frac{100e^{0.48t - \ln 99}}{1 + e^{0.48t - \ln 99}}.$$

Solutions for Section 3.8

1. (a)

TABLE 3.7

x	0	1	2	3	4	5
$y = 3^x$	1	3	9	27	81	243

(b)

TABLE 3.8

x	0	1	2	3	4	5
$y = \log(3^x)$	0	0.477	0.954	1.431	1.909	2.386

The differences between successive terms are constant(≈ 0.477), so the function is linear.

(c)

TABLE 3.9

x	0	1	2	3	4	5
$f(x)$	2	10	50	250	1250	6250

TABLE 3.10

x	0	1	2	3	4	5
$g(x)$	0.301	1	1.699	2.398	3.097	3.796

We see that $f(x)$ is an exponential function (note that it is increasing by a constant growth factor of 5), while $g(x)$ is a linear function with a constant rate of change of 0.699.

(d) The resulting function is linear. If $f(x) = A \cdot B^x$ and $g(x) = \log(A \cdot B^x)$ then

$$g(x) = \log(AB^x)$$
$$= \log A + \log B^x$$
$$= \log A + x \log B$$
$$= b + m \cdot x,$$

where $b = \log A$ and $m = \log B$. Thus, g will be linear.

5. (a) Find the values of $\ln t$ in the table, use linear regression on a calculator or computer with $x = \ln t$ and $y = P$. The line has slope -7.786 and P-intercept 86.28 ($P = -7.786 \ln t + 86.28$). Thus $a = -7.786$ and $b = 86.28$.

(b) Figure 3.8 shows the data points plotted with P against $\ln t$.

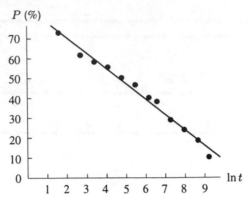

Figure 3.8: Plot of P against $\ln t$ and the line
with slope -7.786 and intercept 86.28

The model seems to fit well.

(c)   The subjects will recognize no words when $P = 0$, that is, when $-7.786 \ln t + 86.28 = 0$. Solving for
t:

$$-7.786 \ln t = -86.28$$
$$\ln t = \frac{86.28}{7.786}$$

Taking both sides to the e power,

$$e^{\ln t} = e^{\frac{86.28}{7.786}}$$
$$t \approx 64{,}954,$$

so $t \approx 45$ days.

The subject recognized all the words when $P = 100$, that is, when $-7.786 \ln t + 86.28 = 100$.
Solving for t:

$$-7.786 \ln t = 13.72$$
$$\ln t = \frac{13.72}{-7.786}$$
$$t \approx 0.17,$$

so $t \approx 0.17$ minutes (≈ 10 seconds) from the start of the experiment.

(d)

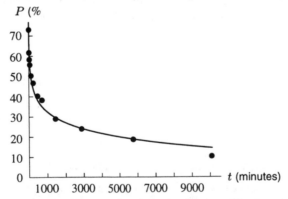

Figure 3.9: The percentage P of words recognized as a
function of t, the time elapsed and the function
$P = -7.786 \ln t + 86.28$

9. (a) Using linear regression we find that the linear function $y = 48 + 0.80x$ gives an excellent fit to the data with a correlation coefficient of $r = 0.9996$.

    (b) To check the fit of an exponential we make a table of x and $\ln y$:

x	30	85	122	157	255	312
$\ln y$	4.248	4.787	4.977	5.165	5.521	5.704

Using linear regression, we find $\ln y = 4.295 + 0.0048x$. Solving for y to put this into exponential form

$$e^{\ln y} = e^{4.295+0.0048x}$$
$$y = e^{4.295}e^{0.0048x}$$
$$y = 73.3e^{0.0048x}$$

which gives a good fit, though not as good as the linear function since $r \approx 0.9728$. Note that since $e^{0.0048} = 1.0048$. We could have written $y = 73.3(1.0048)^x$.

    (c) To try fitting a power function, we make a table of $\ln x$ versus $\ln y$.

$\ln x$	3.401	4.443	4.804	5.056	5.541	5.743
$\ln y$	4.248	4.787	4.977	5.165	5.521	5.704

Running linear regression, we find $\ln y = 2.09 + 0.616 \ln x$. Then

$$e^{\ln y} = e^{2.09+0.616 \ln x}$$
$$e^{\ln y} = e^{2.09}e^{0.616 \ln x}$$

Since $e^{0.616 \ln x} = (e^{\ln x})^{0.616} = x^{0.616}$, we have

$$y = e^{2.09}x^{0.616}$$
$$y = 8.1x^{0.616}$$

which gives a good fit with $r \approx 0.9933$.

    (d) All fits are good. The linear equation gives the best of the three.

13. (a) The function $y = -83 + 61.5x$ gives a superb fit, with correlation coefficient $r = 0.99997$.

    (b) When the power function is plotted for $2 \le x \le 2.05$, it resembles a line. This is true for most of the functions we have studied. If you zoom in close enough on any given point, the function begins to resemble a line. However, for other values of x (say, $x = 3, 4, 5 \ldots$), the fit no longer holds.

Solutions for Chapter 3 Review

1. (a) If a function is linear, then the rate of change is constant. For $Q(t)$,

$$\frac{8.70 - 7.51}{10 - 3} = 0.17.$$

and

$$\frac{9.39 - 8.7}{14 - 10} = 0.17.$$

So this function is very close to linear. Thus, $Q(t) = b + mx$ where $m = 0.17$ as shown above. We solve for b by using the point $(3, 7.51)$.

$$Q(t) = b + 0.17t$$
$$7.51 = b + 0.17(3)$$
$$7.51 - 0.51 = b$$
$$b = 7$$

Therefore, $Q(t) = 7 + 0.17t$.

(b) Testing the rates of change for $R(t)$, we find that

$$\frac{2.61 - 2.32}{9 - 5} = 0.0725$$

and

$$\frac{3.12 - 2.61}{15 - 9} = 0.085,$$

so we know that $R(t)$ is not linear. If $R(t)$ is exponential, then $R(t) = AB^t$, and

$$R(5) = A(B)^5 = 2.32$$

and

$$R(9) = A(B)^9 = 2.61.$$

So

$$\frac{R(9)}{R(5)} = \frac{AB^9}{AB^5} = \frac{2.61}{2.32}$$
$$\frac{B^9}{B^5} = \frac{2.61}{2.32}$$
$$B^4 = \frac{2.61}{2.32}$$
$$B = \left(\frac{2.61}{2.32}\right)^{\frac{1}{4}} \approx 1.030.$$

Since

$$R(15) = A(B)^{15} = 3.12$$
$$\frac{R(15)}{R(9)} = \frac{AB^{15}}{AB^9} = \frac{3.12}{2.61}$$
$$B^6 = \frac{3.12}{2.61}$$
$$B = \left(\frac{3.12}{2.61}\right)^{\frac{1}{6}} \approx 1.030.$$

Since the growth factor, B, is constant, we know that $R(t)$ is an exponential function and that $R = AB^t$. Taking the ratios of $R(5)$ and $R(9)$, we have

$$\frac{R(9)}{R(5)} = \frac{AB^9}{AB^5} = \frac{2.61}{2.32}$$
$$B^4 = 1.125$$
$$B = 1.03$$

So $R(t) = A(1.03)^t$. We now solve for A by using $R(5) = 2.32$

$$R(5) = A(1.03)^5$$
$$2.32 = A(1.03)^5$$
$$A = \frac{2.32}{1.03^5} \approx 2.00$$

Thus, $R(t) = 2.00(1.03)^t$.

(c) Testing rates of change for $S(t)$, we find that

$$\frac{6.72 - 4.35}{12 - 5} = 0.339$$

and

$$\frac{10.02 - 6.72}{16 - 12} = 0.825.$$

Since the rates of change are not the same we know that $S(t)$ is not linear.
Testing for a possible constant growth factor we see that

$$\frac{S(12)}{S(5)} = \frac{AB^{12}}{AB^5} = \frac{6.72}{4.35}$$
$$B^7 = \frac{6.72}{4.35}$$
$$B \approx 1.064$$

and

$$\frac{S(16)}{S(12)} = \frac{AB^{16}}{AB^{12}} = \frac{10.02}{6.72}$$
$$B^4 = \frac{10.02}{6.72}$$
$$B \approx 1.105.$$

Since the growth factors are different, $S(t)$ is not an exponential function.

5. (a) Let p_0 be the price of an item at the beginning of 1980. At the beginning of 1981, its price will be 105.1% of that initial price or $1.051p_0$. At the beginning of 1982, its price will be 106.2% of the price from the year before, that is:

$$\text{Price beginning } 1982 = (1.062)(1.051p_0).$$

By the beginning of 1983, the price will be 103.1% of its price the previous year.

$$\text{Price beginning } 1983 = 1.031(\text{price beginning } 1982)$$
$$= 1.031(1.062)(1.051p_0).$$

Continuing this process,

$$(\text{Price beginning } 1985) = (1.033)(1.047)(1.031)(1.062)(1.051)p_0$$
$$\approx 1.245p_0.$$

So, the cost at the beginning of 1985 is 124.5% of the cost at the beginning of 1980 and the total percent increase is 24.5%.

(b) If r is the average inflation rate for this time period, then $B = 1 + r$ is the factor by which the population on the average grows each year. Using this average growth factor, if the price of an item is initially p_0, at the end of a year its value would be $p_0 B$, at the end of two years it would be $(p_0 B)B = p_0 B^2$, and at the end of five years $p_0 B^5$. According to the answer in part (a), the price at the end of five years is $1.245 p_0$. So

$$p_0 B^5 = 1.245 p_0$$
$$B^5 = 1.245$$
$$B = (1.245)^{\frac{1}{5}} \approx 1.045.$$

If $B = 1.045$, then $r = 0.045$ or 4.5%, the average annual inflation rate.

(c) We assume that the average rate of 4.5% inflation for 1980 through 1984 holds through the beginning of 1990. So, on average, the price of the shower curtain is 104.5% of what it was the previous year for ten years. Then the price of the shower curtain would be $20(1.045)^{10} \approx \$31$.

9. Since this function is exponential, we know $y = AB^x$. We also know that $(-2, 8/9)$ and $(2, 9/2)$ are on the graph of this function, so

$$\frac{8}{9} = AB^{-2}$$

and

$$\frac{9}{2} = AB^2.$$

From these two equations, we can say that

$$\frac{\frac{9}{2}}{\frac{8}{9}} = \frac{AB^2}{AB^{-2}}.$$

Since $(9/2)/(8/9) = 9/2 \cdot 9/8 = 81/16$, we can re-write this equation to be

$$\frac{81}{16} = B^4.$$

Keeping in mind that $B > 0$, we get

$$B = \sqrt[4]{\frac{81}{16}} = \frac{\sqrt[4]{81}}{\sqrt[4]{16}} = \frac{3}{2}.$$

Substituting $B = 3/2$ in $9/2 = AB^2$, we get

$$\frac{9}{2} = A(\frac{3}{2})^2 = \frac{9}{4}A$$
$$A = \frac{\frac{9}{2}}{\frac{9}{4}} = \frac{9}{2} \cdot \frac{4}{9} = \frac{4}{2} = 2.$$

Thus, $y = 2(3/2)^x$.

13. Notice that the x-values are not equally spaced. By finding $\Delta f / \Delta x$ in Table 3.11 we see $f(x)$ is linear, because the rates of change are constant.

TABLE 3.11

x	$f(x)$	$\dfrac{\Delta f(x)}{\Delta(x)}$
0.21	0.03193	
		0.093
0.37	0.04681	
		0.093
0.41	0.05053	
		0.093
0.62	0.07006	
		0.093
0.68	0.07564	

Since $f(x)$ is linear its formula will be $f(x) = b + mx$. From the table, we know $m = 0.093$. Choosing the point $(0.41, 0.05053)$, we have

$$0.05053 = b + 0.093(0.41)$$
$$b = 0.0124.$$

Thus, $f(x) = 0.0124 + 0.093x$.

Alternately, we could have tested $f(x)$ to see if it is exponential with a constant base B. Checking values will show that any proposed B would not remain constant when calculated with different points of function values for $f(x)$ from the table.

Since f is linear, we conclude that g is exponential. Thus, $g(x) = AB^x$. Using two points from above, we have

$$g(0.21) = AB^{0.21} = 3.324896 \qquad g(0.37) = AB^{0.37} = 3.423316.$$

So taking the ratios of $AB^{0.37}$ and AB^{021}, we have

$$\frac{AB^{0.37}}{AB^{0.21}} = \frac{3.423316}{3.324896}$$
$$B^{0.16} \approx 1.0296.$$

So

$$B = (1.0296)^{\frac{1}{0.16}} \approx 1.20.$$

Substituting this value of B in the first equation gives

$$A(1.20)^{0.21} = 3.324896.$$

So

$$A = \frac{3.324896}{1.20^{0.21}}$$
$$A \approx 3.2.$$

Thus, $g(x)$ is approximated by the formula

$$g(x) = 3.2(1.2)^x.$$

17. (a) (i) $0.4 = \frac{2}{5}$, so $\log(2/5) = \log 2 - \log 5 = u - v$

  (ii) $\log 0.25 = \log\left(\frac{1}{4}\right) = \log(2^{-2}) = -2\log 2 = -2u$

  (iii) $\log 40 = \log(2^3 \cdot 5) = \log 2^3 + \log 5 = 3\log 2 + \log 5 = 3u + v$

  (iv) $\log\sqrt{10} = \log 10^{\frac{1}{2}} = \log(2 \cdot 5)^{\frac{1}{2}} = \frac{1}{2}(\log 2 + \log 5) = \frac{1}{2}(u + v)$

  (b)

$$\frac{1}{2}(u + 2v) = \frac{1}{2}(\log 2 + 2\log 5)$$
$$= \frac{1}{2}\log(2 \cdot 5^2)$$
$$= \log\sqrt{50}$$
$$\approx \log\sqrt{49}$$
$$= \log 7.$$

21. The annual growth factors for this investment are 1.27, 1.36, 1.19, 1.44, and 1.57. Thus, the investment increases by a total factor of $(1.27)(1.36)(1.19)(1.44)(1.57) \approx 4.6468$, indicating that the investment is 464.68% of what it had been. If x is the average annual growth factor for this five-year period, this means

$$x^5 = 4.6468$$

and so

$$x = 4.6468^{1/5}$$
$$\approx 1.3597.$$

This tells us that, for each of the five years, the investment has, on average, 135.97% the value of the previous year. It is growing by 35.97% each year. Notice that if we had instead summed these percentages and divided the result by 5, we would have obtained a growth factor of 36.6%. This would be wrong because, as you can check for yourself, 5 years of 36.6% annual growth would result in a total increase of 475.6%, which is not correct.

25. (a)

$$e^{x+3} = 8$$
$$\ln e^{x+3} = \ln 8$$
$$x + 3 = \ln 8$$
$$x = \ln 8 - 3 \approx -0.9206$$

  (b)

$$4(1.12^x) = 5$$
$$1.12^x = \frac{5}{4} = 1.25$$
$$\log 1.12^x = \log 1.25$$
$$x\log 1.12 = \log 1.25$$
$$x = \frac{\log 1.25}{\log 1.12} \approx 1.9690$$

(c)

$$e^{-0.13x} = 4$$
$$\ln e^{-0.13x} = \ln 4$$
$$-0.13x = \ln 4$$
$$x = \frac{\ln 4}{-0.13} \approx -10.6638$$

(d)

$$\log(x - 5) = 2$$
$$x - 5 = 10^2$$
$$x = 10^2 + 5 = 105$$

(e)

$$2\ln(3x) + 5 = 8$$
$$2\ln(3x) = 3$$
$$\ln(3x) = \frac{3}{2}$$
$$3x = e^{\frac{3}{2}}$$
$$x = \frac{e^{\frac{3}{2}}}{3} \approx 1.4939$$

(f)

$$\ln x - \ln(x - 1) = \frac{1}{2}$$
$$\ln\left(\frac{x}{x-1}\right) = \frac{1}{2}$$
$$\frac{x}{x-1} = e^{\frac{1}{2}}$$
$$x = (x - 1)e^{\frac{1}{2}}$$
$$x = xe^{\frac{1}{2}} - e^{\frac{1}{2}}$$
$$e^{\frac{1}{2}} = xe^{\frac{1}{2}} - x$$
$$e^{\frac{1}{2}} = x(e^{\frac{1}{2}} - 1)$$
$$\frac{e^{\frac{1}{2}}}{e^{\frac{1}{2}} - 1} = x$$
$$x \approx 2.5415$$

Note: (g) (h) and (i) can not be solved analytically, so we use graphs to approximate the solutions.

(g) From Figure 3.10 we can see that $y = e^x$ and $y = 3x + 5$ intersect at $(2.534, 12.6)$ and $(-1.599, 0.202)$, so the values of x which satisfy $e^x = 3x + 5$ are $x = 2.534$ or $x = -1.599$. We also see that $y_1 \approx 12.6$ and $y_2 \approx 0.202$.

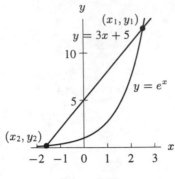

Figure 3.10

(h)   The graphs of $y = 3^x$ and $y = x^3$ are seen in Figure 3.11. It is very hard to see the points of intersection, though $(3, 27)$ would be an immediately obvious choice (substitute 3 for x in each of the formulas). Using technology, we can find a second point of intersection, $(2.478, 15.216)$. So the solutions for $3^x = x^3$ are $x = 3$ or $x = 2.478$.

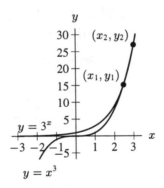

Figure 3.11

Since the points of intersection are very close, it is difficult to see these intersections even by zooming in. So, alternatively, we can find where $y = 3^x - x^3$ crosses the x-axis. See Figure 3.12.

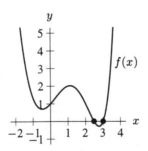

Figure 3.12

(i)   From the graph in Figure 3.13, we see that $y = \ln x$ and $y = -x^2$ intersect at $(0.6529, -0.4263)$, so $x = 0.6529$ is the solution to $\ln x = -x^2$.

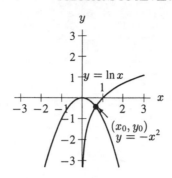

Figure 3.13

29. (a)

$$\text{If } B = 5000(1.06)^t = 5000e^{kt},$$
$$1.06^t = (e^k)^t$$
$$\text{we have } e^k = 1.06.$$

Use the natural log to solve for k,
$$k = \ln(1.06) \approx 0.0583.$$

This means that at a continuous growth rate of 5.83%/year, the account has an effective annual yield of 6%.

(b)

$$7500e^{0.072t} = 7500b^t$$
$$e^{0.072t} = b^t$$
$$e^{0.072} = b$$
$$b \approx 1.0747$$

This means that an account earning 7.2% continuous annual interest has an effective yield of 7.47%.

33. (a) The slope of this line is $m = \frac{y_2 - y_1}{x_2 - x_1} = \frac{3}{2}$. The vertical intercept is 0, thus $y = \frac{3}{2}x$.
    (b) The slope of this line is $\frac{2-1}{0-(-1)} = 1$ and the vertical intercept is 2, thus $\ln y = x + 2$, so $y = e^{x+2}$.
    (c) The slope of this line is $\frac{2-0}{5-0} = 0.4$. The vertical intercept is 0. Thus $\ln y = 0.4x$, and $y = e^{0.4x}$.
    (d) The slope of this line is $\frac{1.7-0}{-1-0} = -1.7$. The vertical intercept is 0. Thus $\ln y = -1.7x$. So $y = e^{-1.7x}$.
    (e) The slope of this line is $\frac{6-0}{4-0} = \frac{3}{2}$. The vertical intercept is 0. Thus $\ln y = \frac{3}{2}\ln x$, and $y = e^{(3/2)\ln x} = e^{\ln(x^{3/2})} = x^{\frac{3}{2}}$.
    (f) The slope of this line is $\frac{2-0}{0-(-3)} = \frac{2}{3}$. The vertical intercept is 2. So $\ln y = 2 + \frac{2}{3}\ln x$, and $y = e^{2+\frac{2}{3}\ln x} = e^2 e^{\frac{2}{3}\ln x} = e^2 e^{\ln(x^{2/3})} = e^2 x^{\frac{2}{3}}$.

CHAPTER FOUR

Solutions for Section 4.1

1.

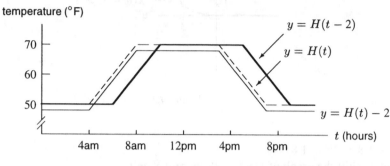

Figure 4.1

(a) The graph of $H(t) - 2$ is the graph of $H(t)$ shifted down by 2 units, or 2°F. Thus, if it uses the $H(t) - 2$ schedule, the company has decided to reduce the temperature in the building by 2°F throughout the day.

(b) The graph of $H(t - 2)$ is the graph of $H(t)$ shifted to the right by 2 units, or 2 hours. Thus, if it uses the $H(t - 2)$ schedule instead of the $H(t)$ schedule, the company has decided to delay all the temperature change by 2 hours.

(c) According to the graph of $H(t)$,

$$H(8) = 70°F,$$
$$H(8) - 2 = 70 - 2 = 68°F,$$
$$H(8 - 2) = H(6) = 60°F.$$

At 8 am, it will be warmest under the $H(t)$ system, namely 70°F. You can also see this in Figure 4.1, where the graph of $H(t)$ is above the other two graphs at $t = 8$.

(d) The $H(t) - 2$ schedule lowers the temperature throughout the day and so will save on heating costs. The $H(t - 2)$ schedule merely shifts the warm period to later in the day.

5. (a) $k(w) - 3 = 3^w - 3$
   To sketch, shift the graph of $k(w) = 3^w$ down 3 units.

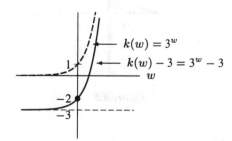

Figure 4.2

(b) $k(w - 3) = 3^{w-3}$
To sketch, shift the graph of $k(w) = 3^w$ to the right by 3 units.

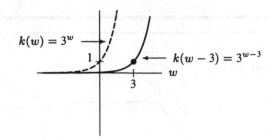

Figure 4.3

(c) $k(w) + 1.8 = 3^w + 1.8$
To sketch, shift the graph of $k(w) = 3^w$ up by 1.8 units.

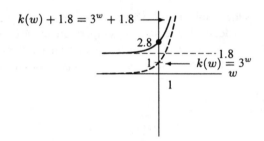

Figure 4.4

(d) $k(w + \sqrt{5}) = 3^{w+\sqrt{5}}$
To sketch, shift the graph of $k(w) = 3^w$ to the left by $\sqrt{5}$ units.

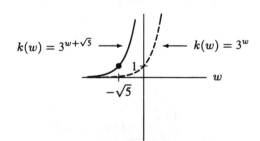

Figure 4.5

(e) $k(w + 2.1) - 1.3 = 3^{w+2.1} - 1.3$
To sketch, shift the graph of $k(w) = 3^w$ to the left by 2.1 units and down 1.3 units.

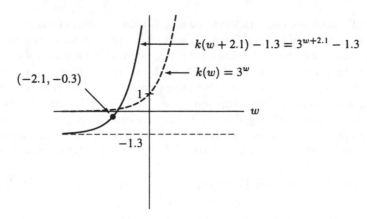

Figure 4.6

(f) $k(w - 1.5) - 0.9 = 3^{w-1.5} - 0.9$
To sketch, shift the graph of $k(w) = 3^w$ to the right by 1.5 units and down by 0.9 units.

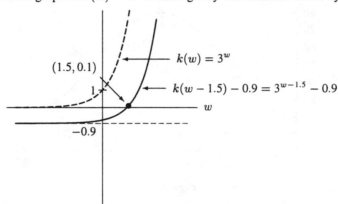

Figure 4.7

9.   (a)   This is an outside change, and thus a vertical change, to $y = |x|$. The graph of $g(x)$ is the graph of $|x|$ shifted upward by 1 unit. See Figure 4.8.
   (b)   This is an inside change, and thus a horizontal change, to $y = |x|$. The graph of $h(x)$ is the graph of $|x|$ shifted to the left by 1 unit. See Figure 4.9.
   (c)   The graph of $j(x)$ involves two transformations of the graph of $y = |x|$. First, the graph is shifted to the right by 2 units. Next, the graph is shifted up by 3 units. Figure 4.10 shows the result at these two consecutive transformations.

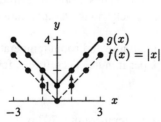

Figure 4.8: Graph of
$g(x) = |x| + 1$,
the graph of $|x|$ shifted up 1 unit

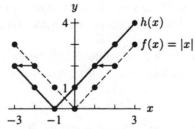

Figure 4.9: Graph of $h(x) = |x + 1|$,
the graph of $|x|$ shifted left 1 unit

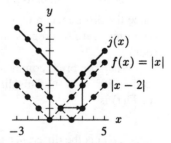

Figure 4.10: Graph of
$j(x) = |x - 2| + 3$,
the graph of $|x|$ shifted right 2
units and up 3 units

13. (a) In looking at the data, we note that the value of $a(t)$ at every value of t is 0.5 greater than the value of $g(t)$ for the same value of t. Thus, for example, $a(0) = g(0) + 0.5$, and in general $a(t) = g(t) + 0.5$.

(b) Studying the data, we note that the data are the same as those in the table for $g(t)$ except that the values in the table of $b(t)$ have been shifted to the left. The value of $b(t)$ at $t = -1.5$ is the value of $g(t)$ at $t = 0$ while the value of $b(t)$ at $t = 0$ is the value of $g(t)$ at $t = 1.5$. Thus, $b(t) = g(t + 1.5)$

(c) In this case, it is easier to first compare $c(t)$ and $b(t)$. In each case, $c(t)$ is 0.3 less than $b(t)$, or $c(t) = b(t) - 0.3$. Since $b(t) = g(t + 1.5)$, we can say that $c(t) = g(t + 1.5) - 0.3$.

(d) As with $b(t)$, $d(t)$ has, in this case, the same values as $g(t)$ except they are shifted to the right by 0.5, so that $d(-1) = g(-1.5) = g(-1 - 0.5), d(0) = g(-0.5) = g(0 - 0.5)$ and $d(1) = g(0.5) = g(1 - 0.5)$. In each case $d(t) = g(t - 0.5)$.

(e) Compare values of $e(t)$ and $d(t)$. For any value of t, $e(t)$ is 1.2 more than $d(t)$. Thus, $e(t) = d(t) + 1.2$. But, $d(t) = g(t - 0.5)$, so $e(t) = g(t - 0.5) + 1.2$.

17. (a) There are many possible graphs, but all should show seasonally-related cycles of temperature increases and decreases.

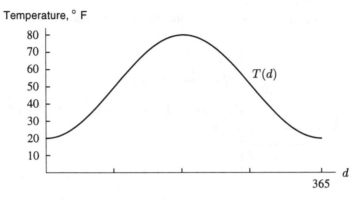

Figure 4.11

(b) While there are a wide variety of correct answers, the value of $T(6)$ is a temperature for a day in early January, $T(100)$ for a day in mid-April, and $T(215)$ for a day in early August. The value for $T(371) = T(365 + 6)$ should be close to that of $T(6)$.

(c) Since there are usually 365 days in a year, $T(d)$ and $T(d + 365)$ represent average temperatures on days which are a year apart.

(d) $T(d + 365)$ is the average temperature on the same day of the year a year earlier. They should be about the same value. Therefore, the graph of $T(d + 365)$ should be about the same as that of $T(d)$.

(e) The graph of $T(d) + 365$ is a shift upward of $T(d)$, by 365 units. It has no significance in practical terms, other than to represent a temperature that is 365° hotter than the average temperature on day d.

21. Since the difference in temperatures decays exponentially, first we find a formula describing that difference over time. Let $D(t)$ represent the difference between the temperature of the brick and the temperature of the room.

When the brick comes out of the kiln, the difference between its temperature and room temperature is $350° - 70° = 280°$. This difference will decay at the constant rate of 3% per minute. Therefore, a formula for $D(t)$ is

$$D(t) = 280(0.97)^t.$$

Since $D(t)$ is the difference between the brick's temperature, $H(t)$, and room temperature, 70°, we have

$$D(t) = H(t) - 70.$$

Add 70 to both sides of the equation so that

$$H(t) = D(t) + 70,$$

Since $D(t) = 280(0.97)^t$,

$$H(t) = 280(0.97)^t + 70.$$

This function, $H(t)$, is *not* exponential because it is not of the form $y = AB^x$. However, since $D(t) = 280(0.97)^t$ *is* exponential, and since

$$H(t) = D(t) + 70,$$

$H(t)$ is a transformation of an exponential function. The graph of $H(t)$ is the graph of $D(t)$ shifted upwards by 70. Figures 4.12 and 4.13 give the graphs of $D(t)$ and $H(t)$ for the first 4 hours, or 240 minutes, after the brick is removed from the kiln—that is, for $0 \le t \le 240$. As you can see, the brick cools off rapidly at first, and then levels off towards 70°, or room temperature, where the graph of $H(t)$ has a horizontal asymptote. Notice that, by shifting the graph of $D(t)$ upwards by 70, the horizontal asymptote is also shifted, resulting in the asymptote at $T = 70$ for $H(t)$.

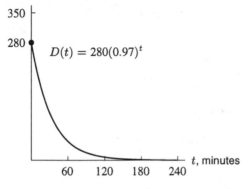

Figure 4.12

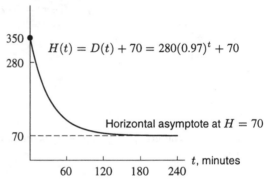

Figure 4.13

Solutions for Section 4.2

1.

TABLE 4.1

p	-3	-2	-1	0	1	2	3
$f(p)$	0	-3	-4	-3	0	5	12

TABLE 4.2

p	-3	-2	-1	0	1	2	3
$g(p)$	12	5	0	-3	-4	-3	0

TABLE 4.3

p	-3	-2	-1	0	1	2	3
$h(p)$	0	3	4	3	0	-5	-12

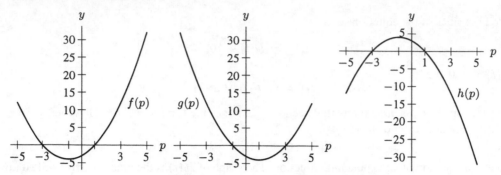

Figure 4.14: Graphs of $f(p)$, $g(p)$, and $h(p)$

Since $g(p) = f(-p)$, the graph of g is a horizontal reflection of the graph of f across the y-axis. Since $h(p) = -f(p)$, the graph of h is a reflection of the graph of f across the p-axis.

5. (a)

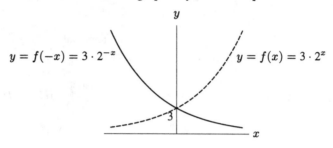

Figure 4.15

The graph of $y = f(-x)$ is the graph of $f(x)$ reflected across the y-axis.

(b)

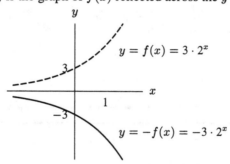

Figure 4.16

The graph of $y = -f(x)$ is the graph of $f(x)$ reflected across the x-axis.

(c)

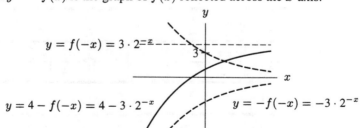

Figure 4.17

The graph of $y = 4 - f(-x) = -f(-x) + 4$ is the graph of $f(x)$ i) reflected across the x-axis, then ii) reflected across the y-axis, then iii) shifted up 4 units.

9. (a) $f(-x) = \sqrt{4 - (-x)^2} = \sqrt{4 - x^2}$.

    (b)

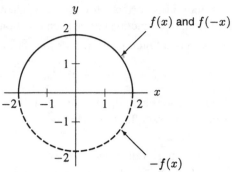

Figure 4.18

    (c) Even.

13.

    (a)

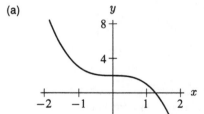

Figure 4.19: $y = -x^3 + 2$

    (b)

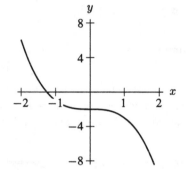

Figure 4.20: $y = -(x^3 + 2)$

    (c) The two functions are not the same.

17.

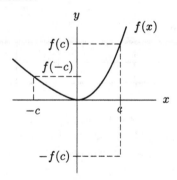

Figure 4.21

21. (a) Since the values of $f(x_0)$ and $f(-x_0)$ are the same, $f(x)$ is symmetric across the y-axis; it is an even function.

    (b) Since the value of $g(-x_0)$ is the opposite of $g(x_0)$, we know that $g(x)$ is symmetric about the origin; it is an odd function.

(c)   The value of $f(-x_0) + g(-x_0)$ is neither $f(x_0) + g(x_0)$ nor $-(f(x_0) + g(x_0))$, so it is not symmetric.

(d)   Note that $f(3 + 1) = f(4) = 13$ and $f(-3 + 1) = f(-2) = -2$. This demonstrates that $f(-x_0 + 1)$ does not equal either $f(x_0 + 1)$ or $-f(x_0 + 1)$.

25.   The argument that $f(x)$ is not odd is correct. However, the statement "something is either even or odd" is false. This function is neither an odd function nor an even function.

29.   To show that $f(x) = x^{1/3}$ is an odd function, we must show that $f(x) = -f(-x)$:

$$-f(-x) = -(-x)^{1/3} = x^{1/3} = f(x).$$

However, not all power functions are odd. The function $f(x) = x^2$ is an even function because $f(x) = f(-x)$ for all x. Another counter-example is $f(x) = \sqrt{x} = x^{1/2}$. This function is not odd because it is not defined for negative values of x.

33.   There is only one such function, and a rather unexciting one at that. Any function with symmetry across the x-axis would look unchanged if you flipped its graph across the x-axis. For any function f, the graph of $y = -f(x)$ is the graph of $y = f(x)$ flipped across the x-axis. Assuming this does not change the appearance of its graph, we have the equation

$$f(x) = -f(x).$$

Adding $f(x)$ to both sides gives

$$2f(x) = 0,$$

or simply

$$f(x) = 0.$$

Thus the only function that is symmetrical across the x-axis is the x-axis itself – that is, the line $y = 0$. If you think about it, you will see that any other curve that is symmetrical across the x-axis would necessarily fail the vertical line test, and would thus not represent the graph of a function.

Solutions for Section 4.3

1.   (a)   To get the table for $f(x)/2$, you need to divide each entry for $f(x)$ by 2 in Table 4.4.

TABLE 4.4

x	-3	-2	-1	0	1	2	3
$f(x)/2$	1	1.5	3.5	$-.5$	-1.5	2	4

(b)   In order to get the table for $-2f(x + 1)$, first get the table for $f(x + 1)$. To do this, note that, if $x = 0$, then $f(x + 1) = f(0 + 1) = f(1) = -3$ and if $x = -4$, then $f(x + 1) = f(-4 + 1) = f(-3) = 2$. Since $f(x)$ is defined for $-3 \le x \le 3$, where x is an integer, then $f(x + 1)$ is defined for $-4 \le x \le 2$.

TABLE 4.5

x	-4	-3	-2	-1	0	1	2
$f(x + 1)$	2	3	7	-1	-3	4	8

Next, multiply each value of $f(x + 1)$ entry by -2.

TABLE 4.6

x	-4	-3	-2	-1	0	1	2
$-2f(x+1)$	-4	-6	-14	2	6	-8	-16

(c) To get the table for $f(x)+5$, you need to add 5 to each entry for $f(x)$ in Table 4.7.

TABLE 4.7

x	-3	-2	-1	0	1	2	3
$f(x)+5$	7	8	12	4	2	9	13

(d) If $x=3$, then $f(x-2)=f(3-2)=f(1)=-3$. Similarly if $x=2$ then $f(x-2)=f(0)=-1$, since $f(x)$ is defined for integral values of x from -3 to 3, $f(x-2)$ is defined for integral values of x, which are two units higher, that is from -1 to 5.

TABLE 4.8

x	-1	0	1	2	3	4	5
$f(x-2)$	2	3	7	-1	-3	4	8

(e) If $x=3$, then $f(-x)=f(-3)=2$, whereas if $x=-3$, then $f(-x)=f(3)=8$. So, to complete the table for $f(-x)$ flip the values of $f(x)$ in Table 4.9 about the origin.

TABLE 4.9

x	-3	-2	-1	0	1	2	3
$f(-x)$	8	4	-3	-1	7	3	2

(f) To get the table for $-f(x)$, take the negative of each value of $f(x)$ from Table 4.10.

TABLE 4.10

x	-3	-2	-1	0	1	2	3
$-f(x)$	-2	-3	-7	1	3	-4	-8

5. (a) Since $y=-f(x)+2$, we first need to reflect the graph of $y=f(x)$ over the x-axis and then shift it upward two units.

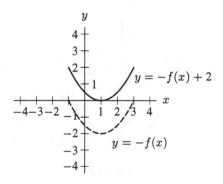

Figure 4.22

(b) We need to stretch the graph of $y = f(x)$ vertically by a factor of 2 in order to get the graph of $y = 2f(x)$.

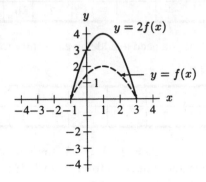

Figure 4.23

(c) In order to get the graph of $y = f(x - 3)$, we will move the graph of $y = f(x)$ to the right by 3 units.

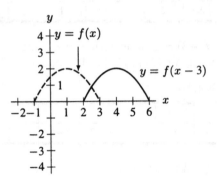

Figure 4.24

(d) To get the graph of $y = -\frac{1}{2}f(x + 1) - 3$, first move the graph of $y = f(x)$ to the left 1 unit, then compress it vertically by a factor of 2. Reflect this new graph over the x-axis and then move the graph down 3 units.

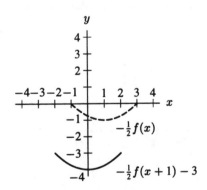

Figure 4.25

9. First let's graph $f(x)$ itself so that we can identify the transformed graphs more easily.

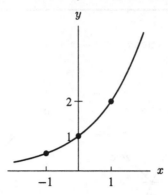

(a) In (a), $f(x)$ has been flipped across the y axis. This means that the new function value for a given x matches the old one for $-x$ and vice versa. Thus, the formula should be $y = f(-x) = 2^{-x}$.

(b) The function in (b) looks just like the function from part (a) except that it has been flipped over the x-axis. In other words, it involves flipping $f(x)$ across both axes. This formula should be $y = -f(-x) = -(2^{-x})$.

(c) Instead of having a horizontal asymptote at the x-axis ($y = 0$), (c) has its asymptote at $y = 1.5$. This means that $f(x)$ has been shifted up 1.5 units, so the formula is $y = f(x) + 1.5 = 2^x + 1.5$.

(d) The curve in (d) is the same shape as $f(x)$, but $x = 0$ corresponds to 5 instead of 1, $x = 1$ corresponds to 10 instead of 2, and so on. So we can see that $f(x)$ has been vertically stretched by a factor of 5. Thus, the formula is $y = 5f(x) = 5 \cdot 2^x$.

13. (a)

TABLE 4.11

x	-4	-3	-2	-1	0	1	2	3	4
$f(-x)$	13	6	1	-2	-3	-2	1	6	13

(b)

TABLE 4.12

x	-4	-3	-2	-1	0	1	2	3	4
$-f(x)$	-13	-6	-1	2	3	2	-1	-6	-13

(c)

TABLE 4.13

x	-4	-3	-2	-1	0	1	2	3	4
$3f(x)$	39	18	3	-6	-9	-6	3	18	39

(d) All three functions are even.

17.

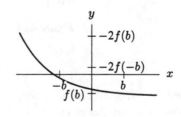

Figure 4.26

Solutions for Section 4.4

1. Figure 4.27 gives graphs of both functions.

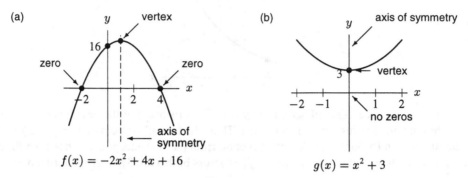

Figure 4.27: Two different quadratic functions.

(a) For f, $a = -2$, $b = 4$, and $c = 16$. The axis of symmetry is the line $x = 1$ and the vertex is at $(1, 18)$. The zeros, or x-intercepts, are at $x = -2$ and $x = 4$. The y-intercept is at $y = 16$.

(b) For g, $a = 1$, $b = 0$, and $c = 3$. Its vertex is at $(0, 3)$, and its axis of symmetry is the y-axis, or the line $x = 0$. This function has no zeros.

5. The function has zeros at $x = -4$ and $x = 5$, and appears quadratic, so it could be of the form $y = a(x + 4)(x - 5)$. Since $y = 40$ when $x = 1$, we know that $y = a(1 + 4)(1 - 5) = -20a = 40$, so $a = -2$. Therefore, $y = -2(x + 4)(x - 5)$.

9. (a)

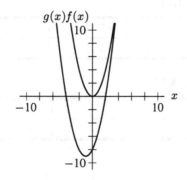

Figure 4.28

On this window we see the expected parabolic shapes of $f(x)$ and $g(x)$. Both $f(x)$ and $g(x)$ are opening upward, so their shapes are similar and the end behaviors are the same. The differences in $f(x)$ and $g(x)$ are apparent at their intercepts. The graph of $f(x)$ has one intercept at $(0, 0)$. The graph of $g(x)$ has x-intercepts at $x = -4$ and $x = 2$, and a y-intercept at $y = -8$.

(b)

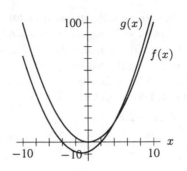

Figure 4.29

As we extend the range to $y = 100$, the difference between the y-intercepts for $f(x)$ and $g(x)$ becomes less significant.

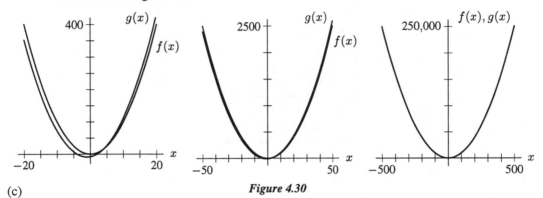

(c)

Figure 4.30

On the window $-20 \leq x \leq 20, -10 \leq y \leq 400$, the graphs are still distinguishable from one another, but all intercepts appear much closer. On the next window, the intercepts appear the same for $f(x)$ and $g(x)$. Only a thickening along the sides of the parabola gives the hint of two functions. On the last window, the graphs appear identical.

13. In order to show that the data in Table 4.35 is accurately modeled by the formula $p(x) = \frac{1}{2}x^2 + \frac{11}{2}x + 8$, we must substitute $x = 0, 1, 2, 3$ (for years 1992-1995) into the formula:

$$p(0) = 8, \ p(1) = 14, \ p(2) = 21, \ p(3) = 29.$$

Our results are consistent with the table. In the year 2002, $x = 10$, and $q(10) = 113$, so the model predicts 113% of schools will have videodisc players in 2002. This is a good model for the period 1992 to 1995, but certainly not around the year 2002 since 113% makes no sense.

17. (a)

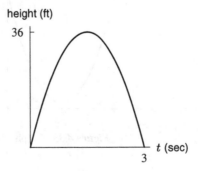

Figure 4.31

(b) To find t when $d(t) = 0$, either use the graph or factor $-16t^2 + 48t$ and set it equal to zero. Factoring yields $-16t^2 + 48t = -16t(t - 3)$, so $d(t) = 0$ when $t = 0$ or $t = 3$. The first time $d(t) = 0$ is at the moment the tomato is being thrown up into the air. The second time is when the tomato hits the ground.

(c) The maximum height occurs on the axis of symmetry, which is halfway between the zeros, at $t = 1.5$. So, the tomato is highest 1.5 seconds after it is thrown.

(d) The maximum height is $d(1.5) = 36$ feet.

21. $x^2 - 1.4x - 3.92 = (x + 1.4)(x - 2.8)$

25. $y = -4cx + x^2 + 4c^2 = x^2 - 4ck + 4c^2 = (x - 2c)^2$. Since $c > 0$, the double zero of this function, $x = 2c$, is positive.

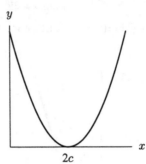

Figure 4.32: $y = -4cx + x^2 + 4c^2$ for $c > 0$

29.

$$y - 12x = 2x^2 + 19$$
$$y = 2x^2 + 12x + 19$$
$$y = 2(x^2 + 6x) + 19$$
$$y = 2(x^2 + 6x + 9) - 18 + 19$$
$$y = 2(x + 3)^2 + 1$$

The graph of this function has an axis of symmetry at $x = -3$ and a vertex at $(-3, 1)$.

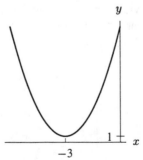

Figure 4.33: Graph of $y - 12x = 2x^2 + 19$

33. Completing the square gives

$$y = (x^2 + 8x + 16) - 16 + 5$$
$$= (x + 4)^2 - 11.$$

Solving the equation $y = 0$ for x gives

$$(x + 4)^2 - 11 = 0$$
$$(x + 4)^2 = 11$$
$$x + 4 = \pm\sqrt{11}$$
$$x = -4 \pm \sqrt{11}.$$

The zeros are $x = -4 + \sqrt{11}$ and $x = -4 - \sqrt{11}$.

37. For a fraction to equal zero, the numerator must equal zero. So, we solve $x^2 - 5mx + 4m^2 = 0$. Since $x^2 - 5mx + 4m^2 = (x - m)(x - 4m)$, we know that the numerator equals zero when $x = 4m$ and when $x = m$. But for $x = m$, the denominator will equal zero as well. So, the fraction is undefined at $x = m$, and the only solution is $x = 4m$.

41. (a)

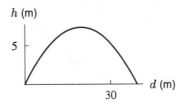

(b) When the ball hits the ground $h = 0$, so $h = 0.75d - 0.01914d^2 = d(0.75 - 0.01914d) = 0$ and we get $d = 0$ or $d \approx 39.185$ m. Since $d = 0$ is the position where the kicker is standing, the ball must hit the ground about 39.2 meters from the point where it is kicked.

(c) The path is parabolic and the maximum height occurs at the vertex, which lies on the axis of symmetry, mid-way between the zeros at $d \approx 19.59$ m. Since $h = 0.75(19.59) - 0.01914(19.59)^2 \approx 7.35$, we know that the ball reaches 7.35 meters above the ground before it begins to fall.

(d) From part (c), the horizontal distance traveled when the ball reaches its maximum height is ≈ 19.59 m.

Solutions for Section 4.5

1. If $x = -2$, then $f(\frac{1}{2}x) = f(\frac{1}{2}(-2)) = f(-1) = 7$, and if $x = 6$, then $f(\frac{1}{2}x) = f(\frac{1}{2} \cdot 6) = f(3) = 8$. In general, $f(\frac{1}{2}x)$ is defined for values of x which are twice the values for which $f(x)$ is defined.

TABLE 4.14

x	-6	-4	-2	0	2	4	6
$f(\frac{1}{2}x)$	2	3	7	-1	-3	4	8

5. The graph of $y = f(2x)$ is a horizontal compression of the graph of $y = f(x)$ by a factor of 2. The graph of $y = f(-\frac{x}{3}) = f(-\frac{1}{3}x)$ is both a horizontal stretch by a factor of 3 and a flip across the y-axis.

(a)

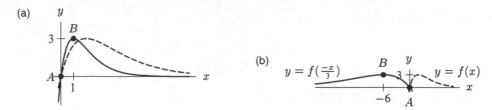

9. (a) To find $g(c)$, locate the point on $y = g(x)$ whose x-coordinate is c. The corresponding y-coordinate is $g(c)$.
   (b) On the y-axis, go up twice the length of $g(c)$ to locate $2g(c)$.
   (c) To find $g(2c)$, you must first find the location of $2c$ on the x-axis. This occurs twice as far from the origin as c. Then find the point on $y = g(x)$ whose x-coordinate is $2c$. The corresponding y-coordinate is $g(2c)$.

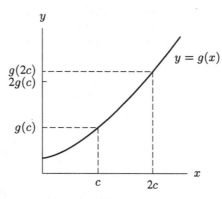

Figure 4.34

Solutions for Chapter 4 Review

1. (a) The input is $2x = 2 \cdot 2 = 4$.
   (b) The input is $\frac{1}{2}x = \frac{1}{2} \cdot 2 = 1$.
   (c) The input is $x + 3 = 2 + 3 = 5$.
   (d) The input is $-x = -2$.

5. (a) $f(10) = 6000$. The total cost for a carpenter to build 10 wooden chairs is $6000.
   (b) $f(30) = 7450$. The total cost for a carpenter to build 30 wooden chairs is $7450.
   (c) $z = 40$. A carpenter can build 40 wooden chairs for $8000.
   (d) $f(0) = 5000$. This is the fixed cost of production, or how much it costs the carpenter to set up before building any chairs.

9. (a) (i) $f(k + 10)$ is how much it costs to produce 10 more than the normal weekly number of chairs.
   (ii) $f(k) + 10$ is 10 dollars more than the cost of a normal week's production.
   (iii) $f(2k)$ is the normal cost of two week's production.
   (iv) $2f(k)$ is twice the normal cost of one week's production (which may be greater than $f(2k)$ since the fixed costs are included twice in $2f(k)$).
   (b) The total amount the carpenter gets will be $1.8f(k)$ plus a five percent sales tax: that is, $1.05(1.8f(k)) = 1.89f(k)$.

13. (a) Sketch varies – should show appropriate seasonal increases/decreases, such as in Figure 4.35.

Figure 4.35: Possible graphs of T and n and p

(b) Freezing is 32°F. If $T(d)$ is the temperature for a particular day, you can determine how far above (or below) freezing $T(d)$ is by subtracting 32 from it. So $n(d) = T(d) - 32$. The sketch of n is the sketch of T shifted 32 units (32°F) downward. See Figure 4.35.

(c) Since low temperatures this year are a week ahead of those of last year, the low temperature on the 100th day of this year, $p(100)$, is the same as the low temperature on the 107th day of last year, $T(107)$. More generally, $p(d) = T(d+7)$. The graph of p is 7 units (7 days) to the left of the graph of T because all low temperatures are occurring seven days earlier. See Figure 4.35

17. (a) After t seconds, the bladder will have contracted $0.25t$ centimeters, so

$$r = f(t) = 12 - 0.25t.$$

(b) The surface area of a sphere of radius r is $4\pi r^2$, so

$$S = g(t) = 4\pi(f(t))^2 = 4\pi(12 - 0.25t)^2$$

(c) We must have r between 12 cm and 2 cm . We know $r = 12$ when $t = 0$. In addition, $r = 2$ when $2 = 12 - 0.25t$, so $t = \dfrac{10}{0.25} = 40$ sec. So the domain of $g(t)$ is $0 \leq t \leq 40$.

(d) We want t such that $g(t) = 100$.

$$4\pi(12 - 0.25t)^2 = 100$$

$$(12 - 0.25t)^2 = \frac{100}{4\pi}$$

$$12 - 0.25t = \pm\frac{10}{\sqrt{4\pi}}$$

$$-0.25t = \pm\frac{10}{2\sqrt{\pi}} - 12$$

$$t = 48 \mp \frac{40}{2\sqrt{\pi}} \approx 37 \quad \text{or} \quad 60 \text{ secs}$$

Since 60 sec is not in the domain, $t \approx 37$ sec. This is the time at which the surface area is 100 cm^2.

(e) We evaluate the function at $t = 0, 10, 20$.

$$g(0) = 4\pi(12)^2 \approx 1810 \quad g(10) = 4\pi(9.5)^2 \approx 1134 \quad g(20) = 4\pi(7)^2 \approx 616$$

Since $1810 - 1134 = 676$ and $1134 - 616 = 518$, the surface area of the bladder will decrease more between $t = 0$ and $t = 10$.

(f) Since $g(0) = 1810$, it's either graph (III) or (IV). We have just shown that the surface area decreases most between $t = 0$ and $t = 10$. Thus, graph (III) could be part of the graph of $S = g(x)$.

CHAPTER FIVE

Solutions for Section 5.1

1.

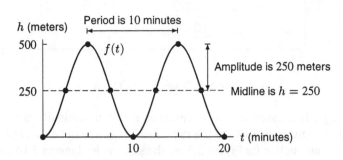

Figure 5.1: Graph of $h = f(t), 0 \leq t \leq 20$

The wheel will complete two full revolutions after 20 minutes, so the function is graphed on the interval $0 \leq t \leq 20$.

5.

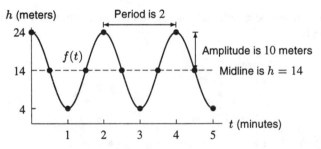

Figure 5.2: Graph of $h = f(t), 0 \leq t \leq 5$

9. At $t = 0$, we see $h = 20$, so you are even with the center of the wheel. This means that your initial position is either three o'clock or nine o'clock. Because initially you are rising, and the wheel is turning counterclockwise, your initial position must have been three o'clock. On the interval $0 \leq t \leq 7$ the wheel completes one and three fourths revolutions. Therefore, if p is the period, we know that

$$\left(1\frac{3}{4}\right)p = \frac{7}{4}p = 7$$

which gives $p = 4$. This means that the ferris wheel takes 4 minutes to complete one full revolution. The minimum value of the function is $h = 5$, which means that you get on and get off of the wheel from a 5 meter platform. The maximum height above the midline is 15 meters, so the wheel's diameter is 30 meters. Notice that the wheel completes a total 2.75 cycles. Since each period is 4 minutes long, you ride the wheel for $4(2.75) = 11$ minutes.

13.

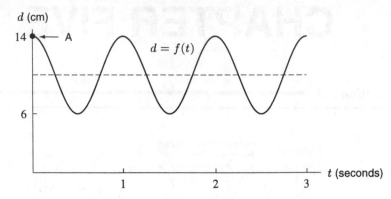

Figure 5.3: Graph of $d = f(t)$ for $0 \leq t \leq 3$

Since the weight is released at $d = 14$ cm when $t = 0$, it is initially at the point in Figure 5.3 labeled A. The weight will begin to oscillate in the same fashion as described by Figures 5.13 and 5.14. Thus, the period, amplitude, and midline for Figure 5.3 are the same as for Figures 5.13 and 5.14.

17.   By plotting the data in Figure 5.4, we can see that the midline is at $h = 4$ (approximately). Since the maximum value is 6.5 and the minimum value is 1.5, we have

$$\text{amplitude} = 4 - 1.5 = 6.5 - 4$$
$$= 2.5.$$

Finally, we can see from the graph that one cycle has been completed from time $t = 0$ to $t = 12$, so the period is 12 seconds.

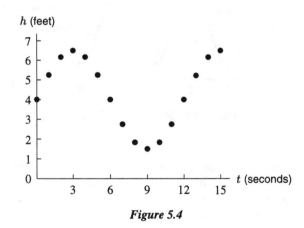

Figure 5.4

21.   Notice that the function is only *approximately* periodic.

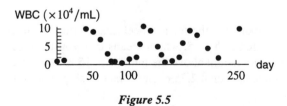

Figure 5.5

The midline is half way between the maximum and minimum WBC values.

$$y = \frac{(10.7 + 0.4)}{2} = 5.55.$$

The amplitude is the difference between the maximum and midline, so $A = 5.15$. The period is the length of time from peak to peak. Measuring between successive peaks gives $p_1 = 120 - 40 = 80$ days; $p_2 = 185 - 120 = 65$ days; $p_3 = 255 - 185 = 70$ days. Using the average of the three periods we get $p \approx 72$ days.

Solutions for Section 5.2

1. The angle $\phi = 420°$ indicates a counterclockwise rotation of the ferris wheel from the 3 o'clock position all the way around once (360°), and then two-thirds of the way back up to the top (an additional 60°). This leaves you in the 1 o'clock position, or at the angle 60°. On the other hand, the angle $\theta = -150°$ indicates a rotation from the 3 o'clock position in the clockwise direction, past the 6 o'clock position and two-thirds of the way up to the 9 o'clock position. This leaves you in the 8 o'clock position, or at the angle 210°. (See Figure 5.6.)

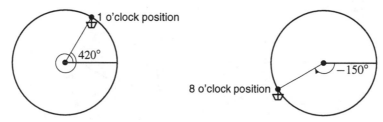

Figure 5.6: The positions and displacements on the ferris wheel described by 420° and $-150°$

5. To locate the points D, E, and F, we mark off their respective angles, $-90°$, $-135°$, and $-225°$, by measuring these angles from the positive x-axis in the clockwise direction. See Figure 5.7.

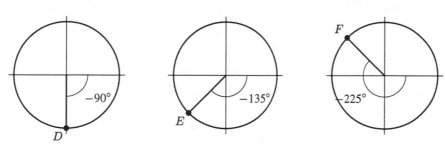

Figure 5.7

9. (a) Since the three panels divide a full rotation — 360° or 2π — into three equal spaces, the angle between each panel is

$$\frac{360°}{3} = 120° \quad \text{or} \quad \frac{2\pi}{3} \text{ radians.}$$

(b)   The angle created by swinging a panel from B to A is equal to half of the angle between each panel, or in other words

$$\frac{120°}{2} = 60° \quad \text{or} \quad \frac{\frac{2\pi}{3}}{2} = \frac{\pi}{3} \text{ radians.}$$

(c)   Note that point B is directly across from point E. Thus the angle of rotation between the two is

$$180° \quad \text{or} \quad \pi \text{ radians.}$$

(d)   A person entering will first push the panel from the initial position C to point B. This is one sixth of the circle. Thus the door rotates an angle of $60°$ or $\frac{\pi}{3}$ radians. The person will then push the panel from B to E covering $180°$ or π degrees. Thus the total angle of rotation of the panel is

$$240° \quad \text{or} \quad \frac{4\pi}{3} \text{ radians.}$$

(e)   From part (d) we know an entry will rotate the door by $\frac{4\pi}{3}$ radians. With three people doing the same rotation, the door will go 4π radians. Thus it will be back to its original position after going around twice. The person leaving then moves the door π radians as found in part (c). Thus the total rotation will be 5π and the panel starting at C will end directly across at F.

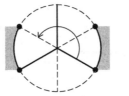 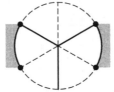

Figure 5.8: The fourth person is exiting the hotel and leaves the door in a new position

13.   Answers vary: The ratio of degree measure to 360 must equal the ratio of radian measure to 2π.

$$\frac{D}{360} = \frac{R}{2\pi} \Rightarrow R = \frac{2\pi D}{360} \Rightarrow R = \frac{\pi}{180}D.$$

Thus $\pi/180$ is the conversion multiplier from degrees to radians.

17.   Like Problem 16, the arc length is equal to the radius times the radian measure, so

$$d = (2)\left[\left(\frac{87}{60}\right)(2\pi)\right] = 5.8\pi \approx 18.22 \text{ inches.}$$

21.   We know $r = 3960$ and $\theta = 1°$. Change θ to radian measure and use $l = r\theta$.

$$l = 3960(1)\left(\frac{\pi}{180}\right) \approx 69.115 \text{ miles.}$$

Solutions for Section 5.3

1. (a) $\sin\theta = 0.6$, $\cos\theta = -0.8$, $\tan\theta = 0.6/(-0.8) = -0.75$
    (b) $\sin\theta = 0.8$, $\cos\theta = -0.6$, $\tan\theta = 0.8/(-0.6) = -4/3$

5. The calculator gives the value $0.707107\ldots$ for both expressions $\sqrt{\frac{1}{2}}$ and $\frac{\sqrt{2}}{2}$. In fact, $\sqrt{\frac{1}{2}} = \frac{\sqrt{2}}{2}$. This is because

$$\sqrt{\frac{1}{2}} = \frac{\sqrt{1}}{\sqrt{2}} = \frac{1}{\sqrt{2}}\frac{\sqrt{2}}{\sqrt{2}} = \frac{\sqrt{2}}{2}.$$

The value 0.7071 is a good approximation of $\sqrt{2}/2$.

9. (a) $\sin(\theta + 360°) = \sin\theta = a$, since the sine function is periodic with a period of $360°$.
    (b) $\sin(\theta + 180°) = -a$. (A point on the unit circle given by the angle $\theta + 180°$ diametrically opposite the point given by the angle θ. So the y-coordinates of these two points are opposite in sign, but equal in magnitude.)
    (c) $\cos(90° - \theta) = \sin\theta = a$. This is most easily seen from the right triangles in Figure 5.9.

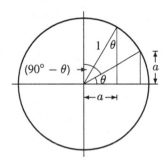

Figure 5.9

(d) $\sin(180° - \theta) = a$. (A point on the unit circle given by the angle $180° - \theta$ has a y-coordinate equal to the y-coordinate of the point on the unit circle given by θ.)
(e) $\sin(360° - \theta) = -a$. (A point on the unit circle given the the angle $360° - \theta$ has a y-coordinate of the same magnitude as the y-coordinate of the point on the unit circle given by θ, but is of opposite sign.)
(f) $\cos(270° - \theta) = -\sin\theta = -a$.

13. The simplest solution is to apply the Pythagorean theorem to the right triangle whose legs are panels and hypotenuse is d. Then $d = \sqrt{1^2 + 1^2} = \sqrt{2}$. An alternate solution uses trigonometry, an approach that works even if the panels do not have a right angle between them. If we consider the circle in Figure 5.42 as describing a unit circle centered at the origin, we can give the coordinates of the points B and C. We do this by first noting that C makes a $\pi/4$ angle with the positive x-axis, and that B makes a $3\pi/4$ angle with the positive x-axis. Then the coordinates of point C are $(\cos\frac{\pi}{4}, \sin\frac{\pi}{4}) = (\sqrt{2}/2, \sqrt{2}/2)$. The coordinates of point B are $(\cos\frac{3\pi}{4}, \sin\frac{3\pi}{4}) = (-\sqrt{2}/2, \sqrt{2}/2)$. The difference between the x-values of these coordinates equals d, so $d = \frac{\sqrt{2}}{2} - (-\frac{\sqrt{2}}{2}) = \sqrt{2}$ meters ≈ 1.414 meters.

17. The point-slope formula for a line is $y = y_0 + m(x - x_0)$, where m is the slope and (x_0, y_0) is a point on the line. Here the slope of line l is $(\sin\theta)/(\cos\theta) = \tan\theta$. Thus, $y = y_0 + (\tan\theta)(x - x_0)$, where (x_0, y_0) is a point on the line.

21.

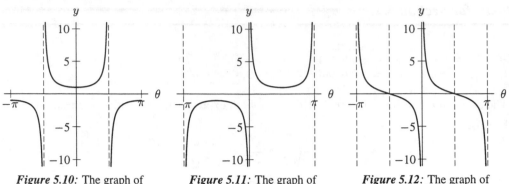

Figure 5.10: The graph of $y = \sec\theta$

Figure 5.11: The graph of $y = \csc\theta$

Figure 5.12: The graph of $y = \cot\theta$

Since the secant and cosecant functions are reciprocals of the cosine and sine function, respectively, they have period 2π. The cotangent function is the reciprocal of the tangent function so its period will be π. As expected, each of these functions tends toward infinity at points where the reciprocal function approaches zero. Thus secant, cosecant, and cotangent all have periodic vertical asymptotes. Each of these functions is positive on the same intervals where the reciprocal function is positive, and each is negative on the intervals where the reciprocal function is negative. The cotangent function has zeros where the values of $\tan\theta$ approach infinity.

25. (a) Yes. Since $\sin t$ repeats periodically, the input for f will repeat periodically, and thus so will the output of f, which makes $f(\sin t)$ periodic. In symbols, if we have $h(t) = f(\sin t)$, then

$$h(t + 2\pi) = f(\sin(t + 2\pi))$$
$$= f(\sin t)$$
$$= h(t).$$

So h is periodic.

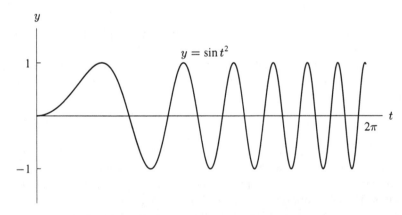

Figure 5.13: The function $y = \sin(t^2)$ is not periodic.

(b) No. For example, $y = \sin(t^2)$ is not periodic, as is clear from Figure 5.13. Although the graph does oscillate up and down, the time from peak to peak is shorter as time increases. Thus the "period" is not constant, which is necessary for a periodic function.

Solutions for Section 5.4

1. $f(t) = 250 + 250 \sin \left(\dfrac{\pi}{5}t - \dfrac{\pi}{2} \right)$

5. $f(t) = 14 + 10 \sin \left(\pi t + \dfrac{\pi}{2} \right)$

9. $f(t) = 20 + 15 \sin \left(\dfrac{\pi}{2}t + \dfrac{\pi}{2} \right)$

Solutions for Section 5.5

1. (a) The function $y = \sin(-t)$ is periodic, and its period is 2π. The function begins repeating every 2π units, as is clear from its graph. Recall from Chapter 4 that $f(-x)$ is a reflection about the y-axis of the graph of $f(x)$, so the periods for $\sin(t)$ and $\sin(-t)$ are the same.

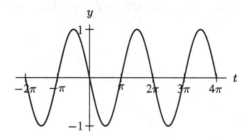

Figure 5.14

(b) The function $y = 4\cos(\pi t)$ is periodic, and its period is 2. This is because when $0 \le t \le 2$, we have $0 \le \pi t \le 2\pi$ and the cosine function has period 2π. Note the amplitude of $4\cos(\pi t)$ is r, but changing the amplitude does not affect the period.

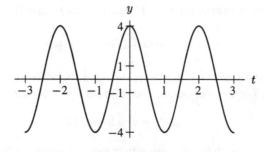

Figure 5.15

(c) The function $y = \sin(t) + t$ is not periodic, because as t gets large, $\sin(t) + t$ gets large as well. In fact, since $\sin(t)$ varies from -1 to 1, y is always between $t - 1$ and $t + 1$. So the values of y cannot repeat.

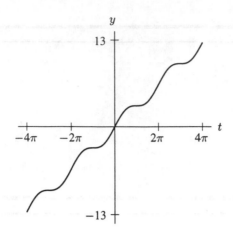

Figure 5.16

(d) In general $f(x)$ and $f(x)+c$ will have the same period if they are periodic. The function $y = \sin(\frac{t}{2})+1$ is periodic, because $\sin(\frac{t}{2})$ is periodic. Since $\sin(t/2)$ completes one cycle for $0 \le t/2 \le 2\pi$, or $0 \le t \le 4\pi$, we see the period of $y = \sin(t/2)+1$ is 4π. See Figure 5.17.

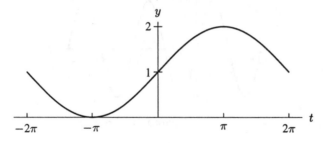

Figure 5.17

5. (a) The graph resembles a sine function that is vertically reflected, horizontally and vertically stretched, and vertically shifted. There is no horizontal shift since the function hits its midline at $\theta = 0$. The midline is halfway between 0 and 4, so it has the equation $y = 2$. The amplitude is 2. Since we see 9 is $\frac{3}{4}$ of the length of a cycle, the period is 12. Hence $B = 2\pi/(\text{period}) = \pi/6$, and so

$$y = -2\sin\left(\frac{\pi}{6}\theta\right) + 2.$$

(b) The graph is a horizontally and vertically compressed sine function. The midline is $y = 0$. The amplitude is 0.8. We see that $\pi/7 = $ two periods, so the period is $\pi/14$. Hence $B = 2\pi/(\text{period}) = 28$, and so

$$y = 0.8\sin(28\theta).$$

(c) The graph is a horizontally stretched cosine function that is vertically shifted down 4 units, but is not horizontally shifted. The midline $y = -4$ is given. The amplitude is 1. The period is 13, so $B = 2\pi/13$. Thus

$$y = \cos\left(\frac{2\pi}{13}\theta\right) - 4.$$

9. We can sketch these graphs using a calculator or computer. Figure 5.18 gives a graph of $y = \sin\theta$ together with the graphs of $y = 2\sin\theta$ and $y = -\frac{1}{2}\sin\theta$ where θ is in radians and $0 \le \theta \le 2\pi$.

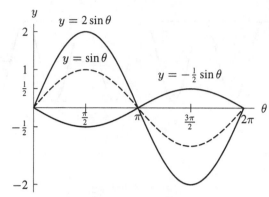

Figure 5.18: The graphs of $y = \sin\theta, y = 2\sin\theta$, and $y = -\frac{1}{2}\sin\theta$ all have different amplitudes.

These graphs are similar but not the same. The amplitude of $y = 2\sin\theta$ is 2 and the amplitude of $y = -\frac{1}{2}\sin\theta$ is $\frac{1}{2}$. The graph of $y = -\frac{1}{2}\sin\theta$ is "upside-down" relative to the other two graphs. These observations are consistent with the fact that the constant A in the equation

$$y = A\sin\theta$$

may result in a vertical stretching or shrinking and/or a reflection over the x-axis.

13. The data given describes a trigonometric function shifted vertically because all the $g(x)$ values are greater than 1. Since the maximum is approximately 3 and the minimum approximately 1, the midline value is 2. We choose the sine function over the cosine function because the data tells us that at $x = 0$ the function takes on its midline value, and then increases. Thus our function will be of the form

$$g(x) = A\sin(Bx) + D.$$

We know that A represents the amplitude, D represents the vertically-shift, and the period is $2\pi/B$.

We've already noted the midline value is $D \approx 2$. This means $A = \text{max} - D = 1$. We also note that the function completes a full cycle after 1 unit. Thus

$$B = \frac{2\pi}{\text{period}} = 2\pi.$$

Thus

$$g(x) = \sin(2\pi x) + 2.$$

17. (a) This function has an amplitude of 3 and a period 1, and resembles a sine graph. Thus $y = 3f(x)$.
    (b) This function has an amplitude of 2 and a period of 3, and resembles vertically reflected cosine graph. Thus $y = -2g(x/3)$.
    (c) This function has an amplitude of 1 and a period of 0.5, and resembles an inverted sine graph. Thus $y = -f(2x)$.
    (d) This function has an amplitude of 2 and a period of 1 and a midline of $y = -3$, and resembles a cosine graph. Thus $y = 2g(x) - 3$.

21.   (a)

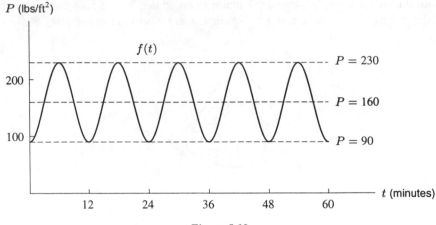

Figure 5.19

This function is vertically reflected cosine function which has been vertically shifted. Thus the function for this equation will be of the form

$$P = f(t) = -A\cos(Bt) + D.$$

(b)   The midline value is $D = (90 + 230)/2 = 160$.
The amplitude is $|A| = 230 - 160 = 70$.
A complete oscillation is made each 12 minutes, so the period is 12. This means $B = 2\pi/12 = \pi/6$.
Thus $P = f(t) = -70\cos(\frac{\pi}{6}t) + 160$.

(c)   Graphing $P = f(t)$ on a calculator for $0 \le t \le 2$ and $90 \le P \le 230$, we see that $P = f(t)$ first equals 115 when $t \approx 1.67$ minutes.

25.

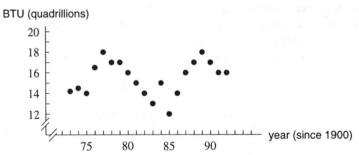

Figure 5.20: U.S. Imports of Petroleum

We start with a sine function of the form

$$f(t) = A\sin(Bt + C) + D.$$

Since the maximum here is 18 and the minimum is 12, the midline value is $D = (18 + 12)/2 = 15$. The amplitude is then $A = 18 - 15 = 3$. The period, measured peak to peak, is 12. So $B = 2\pi/12 = \pi/6$. Lastly, to calculate C, we remember that C/B is the horizontal shift. Our data are close to the midline value for $t \approx 74$, whereas $\sin t$ is at its midline value for $t = 0$. So a (rightward) shift of $C/B = -74$ is needed. This means

$$
\begin{aligned}
C &= -74B \\
&= -74\frac{\pi}{6} \\
&\approx -38.75.
\end{aligned}
$$

So our final equation is

$$f(t) = 3 \sin \left(\frac{\pi}{6} t - 38.75 \right) + 15.$$

We can check our formula by graphing it and seeing how close it comes to the data points. See Figure 5.21.

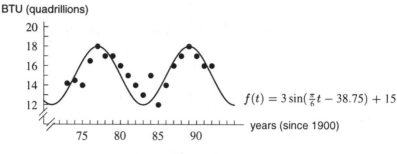

Figure 5.21

Solutions for Section 5.6

1. (a) The period is 12, meaning that the fox population oscillates over a twelve-month period. The midline, $F = 350$, is the average fox population over the year. The amplitude is 200, indicating a variation of at most 200 from the midline, or a total variation of 400 from the minimum to the maximum.

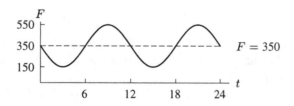

Figure 5.22: The graph of $F = g(t), 0 \le t \le 24$

   (b) The two graphs do support her theory. When the rabbit population is at a minimum, at $t = 0$, the foxes are dying rapidly. To see this, notice that F is decreasing rapidly at $t = 0$. This decrease could be attributed to starvation, which would make sense if the rabbits provided most of the food for the fox population. Furthermore, when the rabbit population is at a maximum, at $t = 6$, the fox population is growing rapidly. This is arguably due to an abundance of food, namely, rabbits.

5. (a) $\arcsin(0.5) = \frac{\pi}{6}$
   (b) $\arccos(-1) = \pi$
   (c) $\arcsin(0.1) \approx 0.1$

9. Using a graphing calculator we graph $y_1 = \sin \theta$ and $y_2 = 0.75$ and look for intersections where $0 \le \theta \le \pi$. We find $\theta \approx 0.848$ and $\theta \approx 2.29$. In exact form this is $\theta = \sin^{-1}(3/4)$ and $\theta = \pi - \sin^{-1}(3/4)$.

13. Graphically, we will only solve part (a). Parts (b)–(g) are solved in much the same way.

(a)

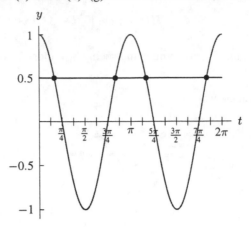

Figure 5.23

From the graph we can see that the solutions lie on the intervals $\frac{\pi}{8} < t < \frac{\pi}{4}$, $\frac{3\pi}{4} < t < \frac{7\pi}{8}$, $\frac{9\pi}{8} < t < \frac{5\pi}{4}$ and $\frac{7\pi}{4} < t < \frac{15\pi}{8}$. Using the trace mode on a calculator, we can find approximate solutions $t_1 = 0.52$, $t_2 = 2.62$, $t_3 = 3.67$ and $t_4 = 5.76$.

For a more precise answer we proceed to solve $\cos(2t) = \frac{1}{2}$ algebraically. The first step is $2t = \arccos(1/2)$. One solution is $2t = \pi/3$. But $2t = 5\pi/3$, $7\pi/3$, and $11\pi/3$ are also angles that have a cosine of $1/2$. Thus $t = \pi/6$, $5\pi/6$, $7\pi/6$, and $11\pi/6$ are the solutions between 0 and 2π.

(b) To solve

$$\tan t = \frac{1}{\tan t}$$

we multiply both sides of the equation by $\tan t$. We must keep in mind that this only makes sense if $\tan t \neq 0$, which is implicitly assumed because of the right hand side of the equation. Multiplication gives us

$$\tan^2 t = 1 \qquad \text{or} \qquad \tan t = \pm 1.$$

This means $t = \arctan(\pm 1) = \pm\pi/4$. There are other angles that have a tan of ± 1, namely $\pm 3\pi/4$. So $t = \pi/4$, $3\pi/4$, $5\pi/4$, and $7\pi/4$ are the solutions in the interval from 0 to 2π.

(c) We factor out the $\cos t$:

$$0 = 2\sin t \cos t - \cos t = \cos t(2\sin t - 1).$$

So solutions occur either when $\cos t = 0$ or $2\sin t - 1 = 0$. The condition $\cos t = 0$ has solutions $\pi/2$ and $3\pi/2$. The condition $2\sin t - 1 = 0$ has solution $t = \arcsin(1/2) = \pi/6$, and also $t = \pi - \pi/6 = 5\pi/6$. Thus the solutions to the original problem are

$$t = \frac{\pi}{2}, \frac{3\pi}{2}, \frac{\pi}{6} \text{ and } \frac{5\pi}{6}.$$

(d) To solve $3\cos^2 t = \sin^2 t$ we divide both sides by $\cos^2 t$ and rewrite the problem as

$$3 = \tan^2 t \qquad \text{or} \qquad \tan t = \pm\sqrt{3}.$$

Using the inverse tangent and reference angles we find that the solutions occur at the points

$$t = \frac{\pi}{3}, \frac{4\pi}{3}, \frac{2\pi}{3} \text{ and } \frac{5\pi}{3}.$$

17. (a) True; divide by e^t, which is never negative, and obtain $\tan t = 2/3$.
     (b) True; graph $y_1 = \cos^{-1}(\sin(\cos^{-1} x))$ for $-1 \le x \le 1$.
     (c) True; we need $a + b \sin t$ to be positive for the natural logarithm to be defined. Since $\sin t \ge -1$, we have $a + b \sin t \ge a - b \ge 0$.
     (d) False; graph $y_1 = \cos(e^t) + \sin(e^t)$ and $y_2 = \sin 1$ to see this is false. For example, if we let $x = 0$ we get

     $$\cos(e^x) + \sin(e^x) = \cos(e^0) + \sin(e^0)$$
     $$= \cos(1) + \sin(1)$$
     $$> \sin(1).$$

Solutions for Section 5.7

1. (a) We can rewrite the equation as follows

     $$0 = \cos 2\theta + \cos \theta = 2\cos^2 \theta - 1 + \cos \theta.$$

     Factoring we get
     $$(2\cos \theta - 1)(\cos \theta + 1) = 0.$$

     Thus the solutions occur when $\cos \theta = -1$ or $\cos \theta = \frac{1}{2}$. These are special values of cosine. If $\cos \theta = -1$ then we have $\theta = 180°$. If $\cos \theta = \frac{1}{2}$ we have $\theta = 60°$ or $300°$. Thus the solutions are

     $$\theta = 60°, \ 180°, \text{ and } 300°.$$

     (b) Using the Pythagorean identity we can substitute $\cos^2 \theta = 1 - \sin^2 \theta$ and get

     $$2(1 - \sin^2 \theta) = 3\sin \theta + 3.$$

     This gives
     $$-2\sin^2 \theta - 3\sin \theta - 1 = 0.$$

     Factoring we get
     $$-2\sin^2 \theta - 3\sin \theta - 1 = -(2\sin \theta + 1)(\sin \theta + 1) = 0.$$

     Thus the solutions occur when $\sin \theta = -\frac{1}{2}$ or when $\sin \theta = -1$. Again knowing special angles allows us to say if $\sin \theta = -\frac{1}{2}$, we have

     $$\theta = \frac{7\pi}{6} \quad \text{and} \quad \frac{11\pi}{6}.$$

     If $\sin \theta = -1$ we have
     $$\theta = \frac{3\pi}{2}.$$

5.

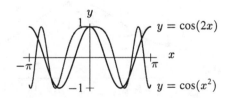

Figure 5.24

Both functions are symmetric about the y-axis. (They are even functions.) They are both equal to one when $x = 0$. They both have an amplitude of one. However, $\cos(2x)$ is periodic, while $\cos(x^2)$ is not.

9. We will use the identity $\cos(2x) = 2\cos^2 x - 1$, where x will be 2θ.

$$\cos 4\theta = \cos(2x)$$
$$= 2\cos^2 x - 1 \quad \text{(using the identity for } \cos(2x))$$
$$= 2(2\cos^2 \theta - 1)^2 - 1 \quad \text{(using the identity for } \cos(2\theta))$$

13. (a) By the Pythagorean theorem, the side adjacent to θ has length $\sqrt{1 - y^2}$. So

$$\cos \theta = \sqrt{1 - y^2}/1 = \sqrt{1 - y^2}.$$

(b) Since $\sin \theta = y/1$, we have

$$\tan \theta = \frac{y}{\sqrt{1 - y^2}}.$$

(c) Using the double angle formula,

$$\cos(2\theta) = 1 - 2\sin^2 \theta = 1 - 2y^2.$$

(d) Supplementary angles have equal sines:

$$\sin(\pi - \theta) = \sin \theta = y.$$

(e) Since $\cos(\pi/2 - \theta) = y$, we have $\sin(\cos^{-1}(y)) = \sin(\pi/2 - \theta) = \sqrt{1 - y^2}$. So

$$\sin^2(\cos^{-1}(y)) = 1 - y^2.$$

Solutions for Section 5.8

1. (a) We know that the minimum of $f(t)$ is 40 and that the maximum is 90. Thus the midline height is

$$\frac{90 + 40}{2} = 65$$

and the amplitude is

$$90 - 65 = 25.$$

We also know that $f(t)$ has a period of 24 hours and is at a minimum when $t = 0$. Thus a formula for $f(t)$ is

$$f(t) = 65 - 25\cos\left(\frac{\pi}{12}t\right).$$

(b) The amplitude is 30. The period is 24 hours, so the pattern repeats itself each day. The midline value is 80. So $g(t)$ goes up to a maximum of $80 + 30 = 110$ and down to a low of $80 - 30 = 50$ megawatts.

(c)

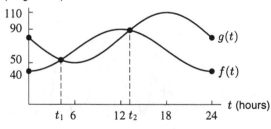

Figure 5.25

From the graph we can find that $t_1 \approx 4.2$ and $t_2 \approx 13.2$. Thus, the power required in both cities is the same at approximately 4 AM and 1 PM.

(d)

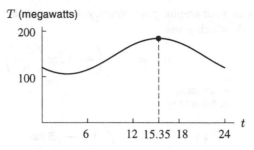

Figure 5.26: $h(t) = f(t) + g(t)$

The function $h(t)$ tells us the total amount of electricity required by both cities at a particular time of day. Using the trace key on a calculator, we find the maximum occurs at $t = 15.35$. So at 3:20 PM each day the most total power will be needed.

5.

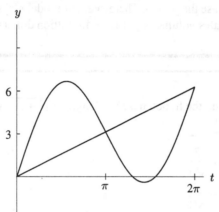

Figure 5.27

For the graphs to intersect, $t + 5 \sin t = t$. So $\sin t = 0$, or $t = $ any multiple of π.

9. (a) Types of video games are trendy for a length of time, during which they are extremely popular and sales are high, later followed by a cooling down period as the users become tired of that particular game type. The game players then become interested in a different game type — and so on.

(b) The sales graph does not fit the shape of the sine or cosine curve, and we would have to say that neither of those functions would give us a reasonable model. However from 1979–1989 the graph does have a basic negative cosine shape, but the amplitude varies.

(c) One way to modify the amplitude over time is to multiply the sine (or cosine) function by an exponential function, such as e^{kt}. So we choose a model of the form

$$s(t) = e^{kt}(-a \cos(Ct) + D),$$

where t is the number of years since 1979. Note the $-a$, which is due to the graph looking like an inverted cosine at 1979. The average value starts at about 1.6, and the period appears to be about 6 years. The amplitude is initially about 1.4, which is the distance between the average value of 1.6 and the first peak value of 3.0. This means

$$s(t) = e^{kt}\left(-1.4 \cos\left(\frac{2\pi}{6}t\right) + 1.6\right).$$

By trial and error on your graphing calculator, you can arrive at a value for the parameter k. A reasonable choice is $k = 0.05$, which gives

$$s(t) = e^{0.05t}\left(-1.4\cos\left(\frac{2\pi}{6}t\right) + 1.6\right).$$

(d)

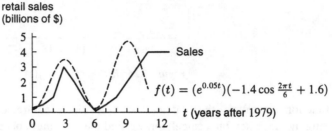

retail sales
(billions of $)

$f(t) = (e^{0.05t})(-1.4\cos\frac{2\pi t}{6} + 1.6)$

Sales

t (years after 1979)

Notice that even though multiplying by the exponential function does increase the amplitude over time, it does not increase the period. Therefore, our model $s(t)$ does not fit the actual curve all that well.

(e)   The predicted 1993 sales volume is $f(14) = 4.6$ billion dollars

Solutions for Section 5.9

1.   By the Pythagorean theorem, the hypotenuse has length $\sqrt{1^2 + 2^2} = \sqrt{5}$

(a) $\tan\theta = \dfrac{\text{opposite}}{\text{adjacent}} = \dfrac{2}{1} = 2.$

(b) $\sin\theta = \dfrac{\text{opposite}}{\text{hypotenuse}} = \dfrac{2}{\sqrt{5}}$

(c) $\cos\theta = \dfrac{\text{adjacent}}{\text{hypotenuse}} = \dfrac{1}{\sqrt{5}}$

5.   Figure 5.28 illustrates this situation.

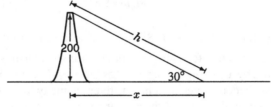

Figure 5.28

We have a right triangle with legs x and 200 and hypotenuse h. Thus,

$$\sin 30° = \frac{200}{h}$$

$$h = \frac{200}{\sin 30°} = \frac{200}{0.5} = 400 \text{ feet.}$$

To find the distance x, we can relate the angle and its opposite and adjacent legs by writing

$$\tan 30° = \frac{200}{x}$$

$$x = \frac{200}{\tan 30°} \approx 346.4 \text{ feet.}$$

We could also write the equation $x^2 + 200^2 = h^2$ and substitute $h = 400$ ft to solve for x.

9. Since $y = \sin \theta$, we can construct the following triangle:

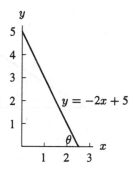

Figure 5.29

The adjacent side, using the Pythagorean theorem, has length $\sqrt{1 - y^2}$. So, $\cos \theta = \dfrac{\text{adj}}{\text{hyp}} = \dfrac{\sqrt{1-y^2}}{1} = \sqrt{1 - y^2}$.

13.

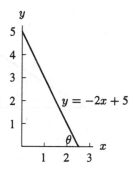

Figure 5.30

The y-intercept of this line is 5; the x-intercept is 2.5. These distances form the legs of the right triangle in Figure 5.30 so $\tan \theta = \dfrac{5}{2.5} = 2$, or $\theta = \tan^{-1}(2) \approx 63.4°$.

17. By the Law of Sines, we have

$$\frac{x}{\sin 100°} = \frac{6}{\sin 18°}$$
$$x = 6 \left(\frac{\sin 100°}{\sin 18°} \right) \approx 19.12.$$

21. (a) In a right triangle $\sin \theta = \frac{\text{opp}}{\text{hyp}}$. Thus $\sin \theta = \frac{3}{7}$. To find $\sin \phi$ we use the Law of Sines:

$$\frac{\sin \phi}{15} = \frac{\sin(20°)}{8}.$$

This implies that

$$\sin \phi = \frac{15 \sin(20°)}{8}.$$

(b) Since $\sin \theta = \dfrac{3}{7}$,

$$\theta = \sin^{-1} \left(\frac{3}{7} \right) \approx 25.4°.$$

This makes sense, as we expect $0° < \theta < 90°$.

For $\sin \phi = \dfrac{15 \sin(20°)}{8}$, there are two solutions

$$\phi = \sin^{-1}\left(\frac{15\sin(20°)}{8}\right) \approx 40° \quad \text{and} \quad \phi = 180° - \sin^{-1}\left(\frac{15\sin(20°)}{8}\right).$$

We choose the second solution $\phi \approx 120°$ since $\phi > 90°$ in Figure 5.121.

25.   The arc is $\frac{30}{360}$ of the circumference. The length of the arc must be

$$\frac{30}{360}10\pi \approx 2.617994 \text{ feet.}$$

Using the Law of Cosines we can solve for the chord length:

$$\text{length of chord } = \sqrt{5^2 + 5^2 - 2(5)(5)\cos 30°} = \sqrt{50 - 50\cos 30°} \approx 2.588191 \text{ feet.}$$

29.   (a)

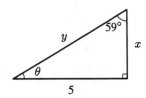

The other angle must be $\theta = 90° - 59° = 31°$.
By definition of the tangent,

$$\tan 59° = \frac{5}{x}$$

$$x = \frac{5}{\tan 59°} \approx 3.0.$$

By definition of the sine,

$$\sin 59° = \frac{5}{y}$$

$$y = \frac{5}{\sin 59°} \approx 5.83.$$

(b)

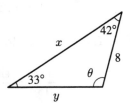

The other angle must be $\theta = 180° - 33° - 42° = 105°$.
By the Law of Sines,

$$\frac{y}{\sin 42°} = \frac{8}{\sin 33°}$$

$$y = 8\left(\frac{\sin 42°}{\sin 33°}\right) \approx 9.83.$$

Again using the Law of Sines,

$$\frac{x}{\sin 105°} = \frac{8}{\sin 33°}$$

$$x = 8\left(\frac{\sin 105°}{\sin 33°}\right) \approx 14.19.$$

(c)

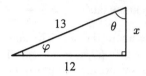

By the Pythagorean theorem, $x = \sqrt{13^2 - 12^2} = 5.$

$$\sin \theta = \frac{12}{13}$$

$$\theta = \sin^{-1}\left(\frac{12}{13}\right)$$

$$\theta \approx 67.38°.$$

Thus $\varphi = 90° - \theta \approx 22.62°.$

33. First draw a radius r as shown in Figure 5.31.

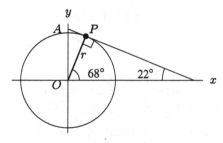

Figure 5.31

Notice $\angle POA = 90° - 68° = 22°$. So $\angle OAP = 90° - 22° = 68°$. Now find r using the Law of Sines
on $\triangle OPA$: $\dfrac{r}{\sin 68°} = \dfrac{4}{\sin 90°}$, which implies that $r = \dfrac{4 \sin 68°}{\sin 90°} \approx 3.71.$
The x and y coordinates of P are

$$x = r \cdot \cos 68° = 3.71 \cos 68° = 1.39$$
$$y = r \cdot \sin 68° = 3.71 \sin 68° = 3.44.$$

Solutions for Chapter 5 Review

1. Answers vary.

5. (a) Each pair is equal, or in other words, $\cos(x) = \cos(-x)$. This follows from the fact that the graph of $y = \cos(x)$ for $x < 0$ is a reflection of the graph of $y = \cos x$ for $x > 0$ about the y-axis (i.e. $\cos x$ is an even function.)

    (b) Each pair is the negative of the other, or in other words, $\sin(-x) = -\sin(x)$. The graph of $y = \sin(x)$ for $x < 0$ is the reflection about the y-axis and then a further reflection in the x-axis of the graph of $y = \sin(x)$ for $x > 0$ (i.e. $y = \sin\theta$ is an odd function.)

    (c) Since $\tan(x) = \dfrac{\sin(x)}{\cos(x)}$, we know $\tan(-x) = \dfrac{\sin(-x)}{\cos(-x)} = \dfrac{-\sin(x)}{\cos(x)} = -\tan(x)$.

9. The data shows the period to be about 0.6 sec. The angular frequency is $b = 2\pi/0.6$. The amplitude is $(\text{high} - \text{low})/2 = (180 - 120)/2 = 60/2 = 30$ cm. The midline is the $(\text{minimum}) + (\text{amplitude}) = 120 + 30 = 150$ cm. It starts in a low position so it is out of phase by $\pi/2$.

$$H(t) = 30\sin((2\pi/0.6)t - \pi/2) + 150.$$

13. (a) $f(t)$ is not defined for in the set $\{\frac{\pi}{2}, \frac{\pi}{2} \pm \pi, \frac{\pi}{2} \pm 2\pi, ...\}$

    (b) $f(t)$ has no limits on its output values: $-\infty < f(t) < +\infty$.

17. We solve

$$3\sin(\pi x - 1) + 1 = 0$$

$$\sin(\pi x - 1) = -\frac{1}{3}$$

$$\pi x - 1 = \arcsin\left(-\frac{1}{3}\right)$$

$$x = \frac{1 + \arcsin\left(-\frac{1}{3}\right)}{\pi}$$

Other solutions are when $\pi x - 1 = \arcsin(-\frac{1}{3}) \pm 2\pi$, and $\pi x - 1 = \arcsin(-\frac{1}{3}) \pm 4\pi$, intersect. These solution are the set $\{\frac{1+\arcsin(-\frac{1}{3})}{\pi}, \frac{1+\arcsin(-\frac{1}{3})\pm 2\pi}{\pi}, \frac{1+\arcsin(-\frac{1}{3})\pm 4\pi}{\pi}, ...\}$. However if $\sin x = k$ is a solution then so is $\pi - x$. This mean that $\pi x - 1 = \pi - \arcsin(-\frac{1}{3})$ is also a solution, and we find $x = \frac{1+\pi-\arcsin(-1/3)}{\pi}$ is a solution. This of course leads to a second infinite set of solutions.

$$\left\{\frac{1 + \pi - \arcsin\left(-\frac{1}{3}\right)}{\pi}, \frac{1 + \pi - \arcsin\left(-\frac{1}{3}\right) \pm 2\pi}{\pi}, \frac{1 + \pi - \arcsin\left(-\frac{1}{3}\right) \pm 4\pi}{\pi}, ...\right\}.$$

21.

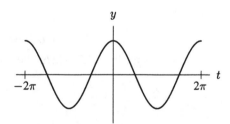

Figure 5.32: Graphs showing $\cos(t) = \sin\left(t + \frac{\pi}{2}\right)$

They are the same graph. This shows us the truth of the identity $\cos t = \sin(t + \frac{\pi}{2})$.

25. (a) The side opposite of angle ϕ has length b and the side adjacent to angle ϕ has length a. Therefore,

$$\sin \phi = \frac{\text{side opposite}}{\text{hypotenuse}} = \frac{b}{c}$$

$$\cos \phi = \frac{\text{side adjacent}}{\text{hypotenuse}} = \frac{a}{c}$$

$$\tan \phi = \frac{\text{side opposite}}{\text{side adjacent}} = \frac{b}{a}.$$

(b)

$$\sin \phi = \frac{\text{side opposite } \phi}{\text{hypotenuse}} = \frac{b}{c},$$

$$\cos \theta = \frac{\text{side adjacent to } \theta}{\text{hypotenuse}} = \frac{b}{c}.$$

Thus $\sin \phi = \cos \theta$. Reversing the roles of ϕ and θ one can show $\cos \phi = \sin \theta$ in exactly the same way.

29. (a)

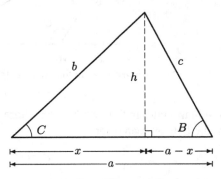

Figure 5.33: Two different right triangles

In Figure 5.33, we see two different right triangles, one whose hypotenuse is b and one whose hypotenuse is c. Applying the Pythagorean theorem to the left triangle, we obtain

$$x^2 + h^2 = b^2.$$

This can be written as

$$h^2 = b^2 - x^2.$$

Applying the Pythagorean theorem to the right triangle gives

$$(a - x)^2 + h^2 = c^2.$$

Substituting $h^2 = b^2 - x^2$ into this equation, we have

$$(a - x)^2 + \underbrace{b^2 - x^2}_{h^2} = c^2.$$

This can be simplified:

$$a^2 - 2ax + x^2 + b^2 - x^2 = c^2$$

$$a^2 + b^2 - 2ax = c^2.$$

From Figure 5.33, we see that, by the definition of cosine,

$$\cos C = \frac{x}{b}.$$

This gives

$$x = b \cos C,$$

which we can substitute into our above equation to obtain

$$a^2 + b^2 - 2a\underbrace{b \cos C}_{x} = c^2,$$

which is the Law of Cosines.

(b)

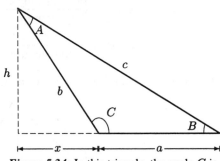

Figure 5.34: In this triangle, the angle C is obtuse

There are two right triangles in Figure 5.34, one whose hypotenuse is b and one whose hypotenuse is c. We have, from the Pythagorean theorem,

$$x^2 + h^2 = b^2,$$

which gives

$$h^2 = b^2 - x^2.$$

We also have

$$(x + a)^2 + h^2 = c^2.$$

Substituting $h^2 = b^2 - x^2$ into this equation gives

$$x^2 + 2ax + a^2 + \underbrace{b^2 - x^2}_{h^2} = c^2,$$

$$a^2 + b^2 + 2ax = c^2.$$

Now, since C is obtuse, $\cos C$ will be negative, and we have

$$\cos C = -\frac{x}{b}$$

which gives

$$x = -b \cos C.$$

Substituting $x = -b \cos C$ into our equation gives

$$a^2 + b^2 - 2a\underbrace{b \cos C}_{x} = c^2,$$

the Law of Cosines.

CHAPTER SIX

1. To construct a table of values for r, we must evaluate $r(0), r(1), \ldots, r(5)$. Starting with $r(0)$, we have

$$r(0) = p(q(0)).$$

Therefore

$$r(0) = p(5) \qquad \text{(because } q(0) = 5\text{)}$$

which, from Table 6.4, gives

$$r(0) = 4.$$

We can repeat this process for $r(1)$:

$$r(1) = p(q(1)) = p(2) = 5.$$

Similarly,

$$r(2) = p(q(2)) = p(3) = 2$$
$$r(3) = p(q(3)) = p(1) = 0$$
$$r(4) = p(q(4)) = p(4) = 3$$
$$r(5) = p(q(5)) = p(8) = \text{ undefined.}$$

These results have been compiled in Table 6.1.

TABLE 6.1

x	0	1	2	3	4	5
$r(x)$	4	5	2	0	3	–

5. To find the simplified formulas, we should use the formula for the innermost function and plug it in to the function that acts on it, replacing it for x's wherever they appear.

   (a) Here, we want to replace each x in the formula for $f(x)$ with the value of $g(x)$, that is, $\dfrac{1}{x-3}$. The
   result is $\left(\dfrac{1}{x-3}\right)^2 + 1 = \dfrac{1}{x^2 - 6x + 9} + 1 = \dfrac{1 + x^2 - 6x + 9}{x^2 - 6x + 9} = \dfrac{x^2 - 6x + 10}{x^2 - 6x + 9}$.

   (b) Similarly, we replace the x's that appear in the formula for $g(x)$ with $x^2 + 1$, the expression for $f(x)$.
   This gives $\dfrac{1}{(x^2 + 1) - 3} = \dfrac{1}{x^2 - 2}$.

   (c) Replacing the x's in the formula for $f(x)$ with $\sqrt{x}$ gives $(\sqrt{x})^2 + 1 = x + 1$.

   (d) Plugging in the expression $x^2 + 1$ for the x term in the formula for $h(x)$ gives $\sqrt{x^2 + 1}$.

(e) Here, we want to take the expression for $g(x)$, namely $\dfrac{1}{x-3}$, and substitute it back into the same expression wherever an x appears. The result is $\dfrac{1}{\frac{1}{x-3}-3}$. We need to simplify the denominator:

$$\frac{1}{x-3}-3 = \frac{1}{x-3} - \frac{3(x-3)}{x-3} = \frac{1-(3x-9)}{x-3} = \frac{10-3x}{x-3}. \text{ So, } \frac{1}{\frac{1}{x-3}-3} = \frac{1}{\frac{10-3x}{x-3}} = \frac{x-3}{10-3x}.$$

(f) Two substitutions have to take place here. The first is to find the value of $f(h(x))$, which we know from part (c) to be $x+1$. Next, we replace each x in the formula for $g(x)$ with an $x+1$. This gives the final result of $\dfrac{1}{(x+1)-3} = \dfrac{1}{x-2}$

9. $k(n(x)) = (n(x))^2 = \left(\dfrac{2x^2}{x+1}\right)^2 = \dfrac{4x^4}{(x+1)^2}$

13.

$$m(m(x)) = \frac{1}{m(x)-1}$$

$$= \frac{1}{\frac{1}{x-1}-1}$$

$$= \frac{1}{\frac{1}{x-1}-\frac{x-1}{x-1}}$$

$$= \frac{1}{\left(\frac{1-(x-1)}{x-1}\right)}$$

$$= \frac{1}{\frac{2-x}{x-1}}$$

$$= \frac{x-1}{2-x}$$

17. (a) We know that $f_7(2) = f(f(f(f(f(f(f(2)))))))$; evaluating this expression would be a tedious process. Let's look for a pattern and see if there is a more elegant solution. Computing small iterations of f gives us:

$$f_1(2) = \frac{1}{2}$$

$$f_2(2) = f(f(2)) = f(1/2) = \frac{1}{1/2} = 2$$

$$f_3(2) = f(f(f(2))) = f(f_2(2)) = f(2) = \frac{1}{2}$$

$$f_4(2) = f(f(f(f(2)))) = f(f_3(2)) = f(1/2) = 2$$

$$f_5(2) = f(f(f(f(f(2))))) = f(f_4(2)) = f(2) = \frac{1}{2}$$

The values of $f_n(x)$ alternate between $\frac{1}{2}$, if n is odd, and 2, if n is even. Since n is odd for $n=7$, we know that $f_7(2) = \dfrac{1}{2}$.

(b)   We can generalize the results in a): If n is odd, then $f_n(x) = \dfrac{1}{x}$ while if n is even, $f_n(x) = x$. Using this generalization, we know that $f_{22}(5) = 5$, which tells us that $f_{23}(f_{22}(5)) = f_{23}(5)$. However, we can use this same generalization to conclude $f_{23}(5) = \dfrac{1}{5}$. So, $f_{23}(f_{22}(5)) = \dfrac{1}{5}$.

21.   These are possible decompositions. There could be others.

(a) $m(x) = u(v(w(x)))$ where $u(x) = \sqrt{x}$, $v(x) = 1 - x$ and $w(x) = x^2$.

(b) $n(x) = u(v(w(x)))$ where $u(x) = \frac{1}{x}$, $v(x) = 1 - x$ and $w(x) = 2x$.

(c) $o(x) = u(v(w(x)))$ where $u(x) = 1 - x$, $v(x) = \sqrt{x}$ and $w(x) = x - 1$.

(d) $p(x) = u(v(w(x)))$ where $u(x) = \sqrt[3]{x}$, $v(x) = 5 - x$ and $w(x) = \sqrt{x}$.

(e) $q(x) = u(v(w(x)))$ where $u(x) = x^2$, $v(x) = 1 + x$ and $w(x) = \frac{1}{x}$.

(f) $r(x) = u(v(w(x)))$ where $u(x) = \frac{1}{x}$, $v(x) = 1 + x$, and $w(x) = \frac{1}{x+1}$.

25. $m(x) = \dfrac{1}{\sqrt{x}}$

29.   (a) $r(x) = p(q(x)) = p(x - 2) = \dfrac{1}{x-2} + 1 = \dfrac{1}{x-2} + \dfrac{x-2}{x-2} = \dfrac{1+x-2}{x-2} = \dfrac{x-1}{x-2}$.

(b)   Let $s(x) = x + 1$ and $t(x) = \frac{1}{x}$. Then $s(t(x)) = \dfrac{1}{x} + 1 = p(x)$.

(c)

$$p(p(a)) = \dfrac{1}{p(a)} + 1$$

$$= \dfrac{1}{\frac{1}{a} + 1} + 1$$

$$= \dfrac{1}{\frac{1+a}{a}} + 1.$$

Since

$$\dfrac{1}{\frac{1+a}{a}} = 1 \cdot \dfrac{a}{1+a}$$

$$= \dfrac{a}{1+a}.$$

We can say that

$$p(p(a)) = \dfrac{a}{a+1} + 1 = \dfrac{a}{a+1} + \dfrac{a+1}{a+1}$$

$$= \dfrac{2a+1}{a+1}$$

33.   To find the formula for $d(a(x))$ we start on the inside of the d function and replace $a(x)$ by its formula. Thus

$$d(\underbrace{a(x)}_{x+5}) = d(x + 5).$$

Now apply the doubling to what was input, $(x + 5)$, and we have

$$d(x + 5) = 2(x + 5) = 2x + 10.$$

To find the formula for $a(d(x))$, we again start on the inside and double the money.

$$a(\underbrace{d(x)}_{2x}) = a(2x).$$

Now apply the add 5 to what was input, $2x$, and we have

$$a(d(x)) = 2x + 5.$$

It should be clear that $d(a(x))$ is more profitable than $a(d(x))$ no matter what x is to begin with.

Solutions for Section 6.2

1. (a) This is the fare for a ride of 3.5 miles. $C(3.5) \approx \$6.25$.
   (b) This is the number of miles you can travel for \$3.50. Between 1 and 2 miles the increase in cost is \$1.50. Setting up a proportion we have:

   $$\frac{1 \text{ additional mile}}{\$1.50 \text{ additional fare}} = \frac{x \text{ additional miles}}{\$3.50 - \$2.50 \text{ additional fare}}$$

   and $x = 0.67$ miles. Therefore

   $$C^{-1}(\$3.5) \approx 1.67.$$

5. (a) $j(25)$ is the average amount of water (in gallons) required daily by a 25-foot oak. However, $j^{-1}(25)$ is the height of an oak requiring an average of 25 gallons of water per day.
   (b) $j(v) = 50$ means that an oak of height v requires 50 gallons of water daily. This statement can be rewritten $j^{-1}(50) = v$.
   (c) This statement can be written $j(z) = p$, or as $j^{-1}(p) = z$.
   (d) • $j(2z)$ is the amount of water required by a tree that is twice average height.
       • $2j(z)$ is enough water for two oak trees of average height. This expression equals $2p$.
       • $j(z + 10)$ is enough water for an oak tree ten feet taller than average.
       • $j(z) + 10$ is the amount of water required by an oak of average height, plus 10 gallons. Thus, this expression equals $p + 10$.
       • $j^{-1}(2p)$ is the height of an oak requiring $2p$ gallons of water.
       • $j^{-1}(p + 10)$ is the height of an oak requiring $p + 10$ gallons of water.
       • $j^{-1}(p) + 10$ is the height of an oak that is 10 feet taller than average. Thus, this expression equals $z + 10$.

9. (a) Since $f(2) = 3$, $f^{-1}(3) = 2$.
   (b) unknown
   (c) Since $f^{-1}(5) = 4$, $f(4) = 5$.
   (d) $f(f^{-1}(2)) = 2$.

13. (a) $D(5) = 100$. The demand at \$5 per unit would be 100 units per week.
    (b) $D(p) = 500 - 200(p - 3) = 500 - 200p + 600 = 1100 - 200p$.
    (c) Solve for p to get $D^{-1}(q)$. We have $p = \frac{1100 - D(p)}{200}$. Rewriting we get $D^{-1}(q) = \frac{1100 - q}{200}$. Thus

    $$D^{-1}(5) = \frac{1100 - 5}{200} = \frac{1095}{200} = 5.475.$$ When 5 units are demanded the price per unit is \$5.48.
    (d) The slope of $D(p)$ is -200, which means that the demand will go down by 200 when the unit price goes up by \$1.

(e) $p = D^{-1}(400) = 3.5$ so the price should be $3.50.

(f) Revenue at 500 units per week: $500(\$3) = \$1500$. Revenue at 400 units per week: $400(\$3.50) = \$1400$. So it would go up by \$100.

17. Let $y = h(x)$. Then

$$y = 12x^3$$

$$x^3 = \frac{y}{12}$$

$$x = \sqrt[3]{\frac{y}{12}}.$$

So $h^{-1}(x) = \sqrt[3]{\dfrac{x}{12}}$.

21. We have $h(h^{-1}(x)) = x$. Thus, by composition of functions,

$$\frac{\sqrt{h^{-1}(x)}}{\sqrt{h^{-1}(x)}+1} = x.$$

We must solve for $h^{-1}(x)$. Our notation can be simplified by making the algebraic substitution $y = h^{-1}(x)$:

$$\frac{\sqrt{y}}{\sqrt{y}+1} = x.$$

Now, by solving for y, we are solving for $h^{-1}(x)$:

$$\sqrt{y} = x(\sqrt{y}+1)$$

$$\sqrt{y} = x\sqrt{y} + x$$

$$\sqrt{y} - x\sqrt{y} = x$$

$$\sqrt{y}(1-x) = x \qquad \text{(factoring)}$$

$$\sqrt{y} = \frac{x}{1-x}$$

$$y = \left(\frac{x}{1-x}\right)^2.$$

Thus,

$$y = h^{-1}(x) = \left(\frac{x}{1-x}\right)^2.$$

25. (a) Dividing by 7 gives $\sin(3x) = 2/7$. This has solutions

$$3x = \arcsin\left(\frac{2}{7}\right) + \quad \text{any multiple of} \quad 2\pi$$

and

$$3x = \pi - \arcsin\left(\frac{2}{7}\right) + \quad \text{any multiple of} \quad 2\pi.$$

So

$$x = \frac{1}{3}\arcsin\left(\frac{2}{7}\right) \pm k\left(\frac{2\pi}{3}\right) \quad \text{or} \quad \pi - \frac{1}{3}\arcsin\left(\frac{2}{7}\right) \pm k\left(\frac{2\pi}{3}\right),$$

where $k = 0, 1, 2, \ldots$

(b)   We take logarithms to help solve when x is in the exponent:

$$2^{x+5} = 3$$
$$\ln(2^{x+5}) = \ln 3$$
$$(x + 5)\ln 2 = \ln 3$$
$$x = \frac{\ln 3}{\ln 2} - 5.$$

(c)   We raise each side to the $\frac{1}{1.05}$ power:

$$x^{1.05} = 1.09$$
$$x = 1.09^{\frac{1}{1.05}}.$$

(d)   We take the exponential function to both sides since the exponential function is the inverse of logarithm:

$$\ln(x + 3) = 1.8$$
$$x + 3 = e^{1.8}$$
$$x = e^{1.8} - 3.$$

(e)   Multiplying by the denominator gives:

$$\frac{2x + 3}{x + 3} = 8$$
$$2x + 3 = 8x + 24$$
$$-21 = 6x$$
$$x = -\frac{21}{6} = -\frac{7}{2}.$$

(f)   Squaring eliminates square roots:

$$\sqrt{x + \sqrt{x}} = 3$$
$$x + \sqrt{x} = 9$$
$$\sqrt{x} = 9 - x \quad (\text{so} \quad x \le 9)$$
$$x = (9 - x)^2 = 81 - 18x + x^2.$$

So $x^2 - 19x + 81 = 0$. The quadratic formula gives the solutions

$$x = \frac{19 \pm \sqrt{37}}{2}.$$

But only one solution is less than 9: $x = \dfrac{19 - \sqrt{37}}{2}$. The other solution fails to satisfy the original equation.

Solutions for Section 6.3

1.

$$h(k(t)) = \sqrt{2\left(\frac{t^2}{2}\right)} = \sqrt{t^2} = t$$

$$k(h(x)) = \frac{(\sqrt{2x})^2}{2} = \frac{2x}{2} = x$$

5. One way to verify that these functions are indeed inverses is to make sure they satisfy the identities $f(f^{-1}(x)) = x$ and $f^{-1}(f(x)) = x$.

$$f(f^{-1}(x)) = 1 + 7\left(\sqrt[3]{\frac{x-1}{7}}\right)^3$$

$$= 1 + 7\left(\frac{x-1}{7}\right)$$

$$= 1 + (x - 1)$$

$$= x.$$

Also,

$$f^{-1}(f(x)) = \sqrt[3]{\frac{1 + 7x^3 - 1}{7}}$$

$$= \sqrt[3]{x^3}$$

$$= x.$$

Thus, $f^{-1}(x) = \sqrt[3]{\dfrac{x-1}{7}}$.

9. (a) $C(0)$ is the concentration of alcohol in the 100 ml solution after 0 ml of alcohol is removed. Thus, $C(0) = 99\%$.
   (b) Note that there are initially 99 ml of alcohol and 1 ml of water.

$$C(x) = \frac{\text{Concentration of alcohol}}{\text{after removing } x \text{ ml}} = \frac{\text{Amount of alcohol remaining}}{\text{Amount of solution remaining}}$$

$$= \frac{\begin{array}{c}\text{Original amount} \\ \text{of alcohol}\end{array} - \begin{array}{c}\text{Amount of} \\ \text{alcohol removed}\end{array}}{\begin{array}{c}\text{Original amount} \\ \text{of solution}\end{array} - \begin{array}{c}\text{Amount of alcohol} \\ \text{removed}\end{array}}$$

$$= \frac{99 - x}{100 - x}.$$

   (c) If $y = C(x)$, then $x = C^{-1}(y)$. We have

$$y = \frac{99 - x}{100 - x}$$

$$y(100 - x) = 99 - x$$
$$100y - xy = 99 - x$$
$$x - xy = 99 - 100y$$
$$x(1 - y) = 99 - 100y$$
$$x = \frac{99 - 100y}{1 - y}.$$

Thus, $C^{-1}(y) = \dfrac{99 - 100y}{1 - y}$.

(d) $C^{-1}(y)$ tells us how much alcohol we need to remove in order to obtain a solution whose concentration is y.

13. We know that

$$A = 3\pi x^3$$

So

$$\frac{A}{3\pi} = x^3$$

$$x = \left(\frac{A}{3\pi}\right)^{1/3}$$

Since

$$B = \frac{x^2}{2\pi}$$

we get

$$B = \frac{\left(\left(\frac{A}{3\pi}\right)^{1/3}\right)^2}{2\pi}$$
$$= \frac{A^{2/3}}{(3\pi)^{2/3}} \cdot \frac{1}{2\pi}$$
$$= \frac{A^{2/3}}{2\pi(3\pi)^{2/3}}$$

17. (a) Start with $x = f(f^{-1}(x))$, and let $y = f^{-1}$ the $x = f(y)$ means

$$x = \frac{3 + 2y}{2 - 5y}$$
$$x(2 - 5y) = 3 + 2y$$
$$2x - 5xy = 3 + 2y$$
$$2x - 3 = 5xy + 2y$$
$$2x - 3 = y(5x + 2)$$
$$y = \frac{2x - 3}{5x + 2},$$

so $y = f^{-1}(x) = \dfrac{2x - 3}{5x + 2}$.

(b)

$$x = \sqrt{\frac{4 - 7y}{4 - y}}$$

$$x^2 = \frac{4 - 7y}{4 - y}$$

$$x^2(4 - y) = 4 - 7y$$

$$4x^2 - xy^2 = 4 - 7y$$

$$4x^2 - 4 = xy^2 - 7y$$

$$4x^2 - 4 = y(x^2 - 7)$$

$$y = \frac{4x^2 - 4}{x^2 - 7},$$

so $y = f^{-1}(x) = \dfrac{4x^2 - 4}{x^2 - 7}$.

(c)

$$x = \frac{1}{9 - \sqrt{y - 4}}$$

$$\frac{1}{x} = 9 - \sqrt{y - 4} \qquad \text{taking reciprocals}$$

$$\sqrt{y - 4} = 9 - \frac{1}{x}$$

$$y - 4 = \left(9 - \frac{1}{x}\right)^2$$

$$y = \left(9 - \frac{1}{x}\right)^2 + 4,$$

so $y = f^{-1}(x) = \left(9 - \dfrac{1}{x}\right)^2 + 4$.

(d)

$$x = \frac{\sqrt{y} + 3}{11 - \sqrt{y}}$$

$$x(11 - \sqrt{y}) = \sqrt{y} + 3$$

$$11x - x\sqrt{y} = \sqrt{y} + 3$$

$$11x - 3 = \sqrt{y} + x\sqrt{y}$$

$$11x - 3 = \sqrt{y}(1 + x)$$

$$\sqrt{y} = \frac{11x - 3}{1 + x}$$

$$y = \left(\frac{11x - 3}{1 + x}\right)^2,$$

so $y = f^{-1}(x) = \left(\dfrac{11x - 3}{1 + x}\right)^2$.

Solutions for Section 6.4

1. Since $h(x) = f(x) + g(x)$, we know that $h(-1) = f(-1) + g(-1) = -4 + 4 = 0$. Similarly, $j(x) = 2f(x)$ tells us that $j(-1) = 2f(-1) = 2(-4) = -8$. Repeat this process for each entry in the table.

TABLE 6.2

x	$h(x)$	$j(x)$	$k(x)$	$m(x)$
-1	0	-8	16	-1
0	0	-2	1	-1
1	2	4	0	0
2	6	10	1	0.2
3	12	16	16	0.5
4	20	22	81	9/11

5. (a) A formula for $h(x)$ would be

$$h(x) = f(x) + g(x).$$

To evaluate $h(x)$ for $x = 3$, we use this equation:

$$h(3) = f(3) + g(3).$$

Since $f(x) = x + 1$, we know that

$$f(3) = 3 + 1 = 4.$$

Likewise, since $g(x) = x^2 - 1$, we know that

$$g(3) = 3^2 - 1 = 9 - 1 = 8.$$

Thus, we have

$$h(3) = 4 + 8 = 12.$$

To find a formula for $h(x)$ in terms of x, we substitute our formulas for $f(x)$ and $g(x)$ into the equation $h(x) = f(x) + g(x)$:

$$h(x) = \underbrace{f(x)}_{x+1} + \underbrace{g(x)}_{x^2-1}$$
$$h(x) = x + 1 + x^2 - 1 = x^2 + x.$$

To check this formula, we use it to evaluate $h(3)$, and see if it gives $h(3) = 12$, which is what we got before. The formula is $h(x) = x^2 + x$, so it gives

$$h(3) = 3^2 + 3 = 9 + 3 = 12.$$

This is the result that we expected.

(b) A formula for $j(x)$ would be

$$j(x) = g(x) - 2f(x).$$

To evaluate $j(x)$ for $x = 3$, we use this equation:

$$j(3) = g(3) - 2f(3).$$

We already know that $g(3) = 8$ and $f(3) = 4$. Thus,

$$j(3) = 8 - 2 \cdot 4 = 8 - 8 = 0.$$

To find a formula for $j(x)$ in terms of x, we again use the formulas for $f(x)$ and $g(x)$:

$$j(x) = \underbrace{g(x)}_{x^2-1} - 2\underbrace{f(x)}_{x+1}$$
$$= (x^2 - 1) - 2(x + 1)$$
$$= x^2 - 1 - 2x - 2$$
$$= x^2 - 2x - 3.$$

We check this formula using the fact that we already know $j(3) = 0$. Since we have $j(x) = x^2 - 2x - 3$,

$$j(3) = 3^2 - 2 \cdot 3 - 3 = 9 - 6 - 3 = 0.$$

This is the result that we expected.

(c)   A formula for $k(x)$ would be

$$k(x) = f(x)g(x).$$

Evaluating $k(3)$, we have

$$k(3) = f(3)g(3) = 4 \cdot 8 = 32.$$

A formula in terms of x for $k(x)$ would be

$$k(x) = \underbrace{f(x)}_{x+1} \cdot \underbrace{g(x)}_{x^2-1}$$
$$= (x + 1)(x^2 - 1)$$
$$= x^3 - x + x^2 - 1$$
$$= x^3 + x^2 - x - 1.$$

To check this formula,

$$k(3) = 3^3 + 3^2 - 3 - 1$$
$$= 27 + 9 - 3 - 1$$
$$= 32,$$

which agrees with what we already knew.

(d)   A formula for $m(x)$ would be

$$m(x) = \frac{g(x)}{f(x)}.$$

Using this formula, we have

$$m(3) = \frac{g(3)}{f(3)} = \frac{8}{4} = 2.$$

To find a formula for $m(x)$ in terms of x, we write

$$m(x) = \frac{g(x)}{f(x)} = \frac{x^2 - 1}{x + 1}$$
$$= \frac{(x + 1)(x - 1)}{(x + 1)}$$
$$= x - 1 \text{ for } x \neq -1$$

We were able to simplify this formula by first factoring the numerator of the fraction $\dfrac{x^2 - 1}{x + 1}$. To check this formula,

$$m(3) = 3 - 1 = 2,$$

which is what we were expecting.

(e) We have

$$n(x) = \left(f(x)\right)^2 - g(x).$$

This means that

$$n(3) = \left(f(3)\right)^2 - g(3)$$
$$= (4)^2 - 8$$
$$= 16 - 8$$
$$= 8.$$

A formula for $n(x)$ in terms of x would be

$$n(x) = (f(x))^2 - g(x)$$
$$= (x + 1)^2 - (x^2 - 1)$$
$$= x^2 + 2x + 1 - x^2 + 1$$
$$= 2x + 2.$$

To check this formula,

$$n(3) = 2 \cdot 3 + 2 = 8,$$

which is what we were expecting.

9. (a)

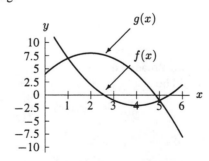

Figure 6.1

(b)

TABLE 6.3

x	0	1	2	3	4	5	6
$f(x)$	14	7	2	−1	−2	−1	2
$g(x)$	4	7	8	7	4	−1	−8
$f(x) - g(x)$	10	0	−6	−8	−6	0	10

(c) See above.

(d) $f(x) = x^2 - 8x + 14$, $g(x) = -x^2 + 4x + 4$,
   $f(x) - g(x) = 2x^2 - 12x + 10$.

(e)

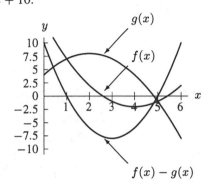

Figure 6.2

13. We can find the revenue function as a product:

$$\text{Revenue} = (\text{\# of customers}) \cdot (\text{price per customer}).$$

At the current price, 50,000 people attend every day. Since 2500 customers will be lost for each \$1 increase in price, the function $n(i)$ giving the number of customers who will attend given i one-dollar price increases, is given by $n(i) = 50{,}000 - 2500i$. The price function $p(i)$ giving the price after i one-dollar price increases is given by $p(i) = 15 + i$. The revenue function $r(i)$ is given by

$$\begin{aligned}
r(i) &= n(i)p(i) \\
&= (50{,}000 - 2500i)(15 + i) \\
&= -2500i^2 + 12{,}500i + 750{,}000 \\
&= -2500(i - 20)(i + 15).
\end{aligned}$$

The graph $r(i)$ is a downward-facing parabola with zeros at $i = -15$ and $i = 20$, so the maximum revenue occurs at $i = 2.5$ which is halfway between the zeros. Thus, to maximize profits the ideal price is $\$15 + 2.5(\$1.00) = \$15 + \$2.50 = \$17.50$.

17. In order to evaluate $h(3)$, we need to express the formula for $h(x)$ in terms of $f(x)$ and $g(x)$. Factoring gives

$$h(x) = C^{2x}(kx^2 + B + 1).$$

Since $g(x) = C^{2x}$ and $f(x) = kx^2 + B$, we can re-write the formula for $h(x)$ as

$$h(x) = g(x) \cdot (f(x) + 1).$$

Thus,

$$h(3) = g(3) \cdot (f(3) + 1)$$
$$= 5(7 + 1)$$
$$= 40.$$

Solutions for Chapter 6 Review

1. (a) $f(2x) = (2x)^2 + (2x) = 4x^2 + 2x$
    (b) $g(x^2) = 2x^2 - 3$
    (c) $h(1 - x) = \dfrac{(1 - x)}{1 - (1 - x)} = \dfrac{1 - x}{x}$
    (d) $(f(x))^2 = (x^2 + x)^2$
    (e) Since $g(g^{-1}(x)) = x$, we have

$$2g^{-1}(x) - 3 = x$$
$$2g^{-1}(x) = x + 3$$
$$g^{-1}(x) = \frac{x + 3}{2}.$$

   (f) $(h(x))^{-1} = \left(\dfrac{x}{1 - x}\right)^{-1} = \dfrac{1 - x}{x}$
   (g) $f(x)g(x) = (x^2 + x)(2x - 3)$
   (h) $h(f(x)) = h(x^2 + x) = \dfrac{x^2 + x}{1 - (x^2 + x)} = \dfrac{x^2 + x}{1 - x^2 - x}$

5.

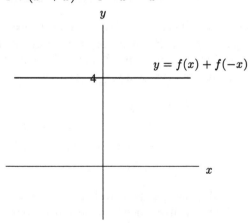

Figure 6.3

   $f(x)$ is linear, and its y-intercept is 2. Thus, a formula for $f(x)$ is given by $f(x) = mx + 2$. This means that the graph of $y = f(x) + f(-x)$ is given by

$$y = f(x) + f(-x) = mx + 2 + m(-x) + 2 = mx - mx + 2 + 2 = 4.$$

This is the horizontal line $y = 4$. The graph is shown in Figure 6.3.

9. (a) Since $P(t)$ is exponential, we know that there are a few ways we could write its formula. One possibility is $P(t) = P_0 b^t$, where b is the annual growth factor. Another is $P(t) = P_0 e^{kt}$ where k is the continuous annual growth rate. If we choose the first, we would have

$$P(t) = P_0 b^t.$$

Since the town triples in size every 7 years, we also know that

$$P(7) = 3P_0.$$

By our formula, we have

$$P(7) = P_0 b^7$$
$$3P_0 = P_0 b^7$$
$$b^7 = 3$$
$$b = 3^{\frac{1}{7}}.$$

Thus, $P(t) = P_0(3^{\frac{1}{7}})^t = P_0(3)^{\frac{t}{7}}$. Since $3^{\frac{1}{7}} \approx 1.17$, we could write $P(t) = P_0(1.17)^t$. If we had used the form $P(t) = P_0 e^{kt}$, we would have found that $P(t) = P_0 e^{0.157t}$.

(b) Since $P(t) = P_0(3^{\frac{1}{7}})^t \approx P_0(1.17)^t$, we know that the town's population increases by about 17% every year.

(c) Letting $x = P(t)$, we solve for t:

$$x = P_0(3)^{\frac{t}{7}}$$
$$3^{\frac{t}{7}} = \frac{x}{P_0}$$
$$\log 3^{\frac{t}{7}} = \log \frac{x}{P_0}$$
$$\left(\frac{t}{7}\right) \log 3 = \log \frac{x}{P_0}$$
$$\frac{t}{7} = \frac{\log \frac{x}{P_0}}{\log 3}$$
$$t = \frac{7 \log \frac{x}{P_0}}{\log 3}.$$

So, $P^{-1}(x) = \frac{7 \log \frac{x}{P_0}}{\log 3} = \frac{7(\log x - \log P_0)}{\log 3}$. $P^{-1}(x)$ is the number of years required for the population to reach x people. Note that there are many different ways to express this formula.

(d) We want to solve $P(t) = 2P_0$. This is equivalent to evaluating $P^{-1}(2P_0)$.

$$P^{-1}(2P_0) = \frac{7 \log \frac{2P_0}{P_0}}{\log 3}$$
$$= \frac{7 \log 2}{\log 3} \approx 4.42 \text{ years.}$$

13.   (a)

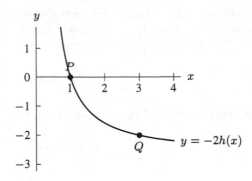

Figure 6.4: $y = -2h(x)$

(b)

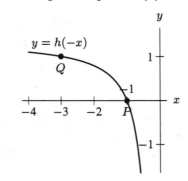

Figure 6.5: $y = h(-x)$

(c)

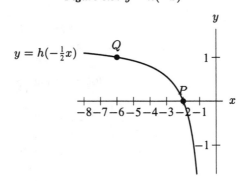

Figure 6.6: $y = h(\frac{-1}{2}x)$

(d)

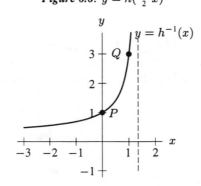

Figure 6.7: $y = h^{-1}(x)$

17. $f(2x) < 2f(x)$

21. The inequality $h(f(x)) < x$ tells us that Space can build fewer than x square feet of office space with the money Ace needs to build x square feet. You get more for your money with Ace.

25. In each case, we will solve the equation $y = f(x)$ for x in order to obtain a formula for $x = f^{-1}(y)$.

(a)

$$y = 3x - 7$$
$$y + 7 = 3x$$
$$\frac{y + 7}{3} = x$$

Therefore,

$$f^{-1}(x) = \frac{x + 7}{3}.$$

(b)

$$y = \frac{1}{x} - 2$$
$$y + 2 = \frac{1}{x}$$
$$x(y + 2) = 1$$
$$x = \frac{1}{y + 2}$$

Therefore,

$$g^{-1}(x) = \frac{1}{x + 2}.$$

(c)

$$y = \frac{2x + 1}{3x - 2}$$
$$y(3x - 2) = 2x + 1$$
$$3xy - 2y = 2x + 1$$
$$3xy - 2x = 2y + 1$$
$$x(3y - 2) = 2y + 1 \quad \text{(factor out an } x\text{)}$$
$$x = \frac{2y + 1}{3y - 2}$$

Therefore,

$$h^{-1}(x) = \frac{2x + 1}{3x - 2}.$$

(d)

$$y = \sqrt{1 + \sqrt{x}}$$
$$y^2 = 1 + \sqrt{x}$$
$$y^2 - 1 = \sqrt{x}$$
$$(y^2 - 1)^2 = x$$

Therefore,

$$j^{-1}(x) = (x^2 - 1)^2.$$

(e)

$$y = \frac{3 - \sqrt{x}}{\sqrt{x} + 2}$$

$$y(\sqrt{x} + 2) = 3 - \sqrt{x}$$

$$y\sqrt{x} + 2y = 3 - \sqrt{x}$$

$$y\sqrt{x} + \sqrt{x} = 3 - 2y$$

$$\sqrt{x}(y + 1) = 3 - 2y \quad \text{(factor out a } \sqrt{x})$$

$$\sqrt{x} = \frac{3 - 2y}{y + 1}$$

$$x = \left(\frac{3 - 2y}{y + 1}\right)^2$$

Therefore,

$$k^{-1}(x) = \left(\frac{3 - 2x}{x + 1}\right)^2.$$

(f)

$$y = \frac{2 - \frac{1}{x}}{3 - \frac{2}{x}}$$

$$y = \frac{2x - 1}{3x - 2} \quad \text{(multiply top and bottom by } x)$$

$$y(3x - 2) = 2x - 1$$

$$3xy - 2y = 2x - 1$$

$$3xy - 2x = 2y - 1$$

$$x(3y - 2) = 2y - 1 \quad \text{(factor out an } x)$$

$$x = \frac{2y - 1}{3y - 2}$$

Thus,

$$l^{-1}(x) = \frac{2x - 1}{3x - 2}.$$

29. (a) Let $y = f(x) = \cos\sqrt{x}$. Solving for x, we have

$$\cos\sqrt{x} = y$$

$$\sqrt{x} = \arccos y$$

$$x = (\arccos y)^2.$$

Replacing y by x and x by $f^{-1}(x)$, we have $f^{-1}(x) = (\arccos x)^2$.

(b) Let $y = g(x) = 2^{\sin x}$. Solving for x gives

$$2^{\sin x} = y$$

$$\ln\left(2^{\sin x}\right) = \ln y$$

$$\ln(2^{\sin x}) = (\sin x)\ln 2 = \ln y$$

$$\sin x = \frac{\ln y}{\ln 2}$$

$$x = \arcsin\left(\frac{\ln y}{\ln 2}\right).$$

Thus $g^{-1}(x) = \arcsin\left(\frac{\ln x}{\ln 2}\right).$

(c) Let $y = h(x) = \sin(2x)\cos(2x)$. Solving for x gives

$$\sin(2x)\cos(2x) = y$$

$$\left(\frac{1}{2}\right)\underbrace{(2\sin(2x)\cos(2x))}_{\sin(2\cdot 2x)} = y$$

$$0.5\sin(4x) = y$$

$$4x = \arcsin(2y)$$

$$x = \frac{\arcsin(2y)}{4}.$$

Thus, $h^{-1}(x) = \dfrac{\arcsin(2x)}{4}$.

(d) Let $y = j(x) = \dfrac{\sin x}{2 - \sin x}$. Solving for x gives

$$\frac{\sin x}{2 - \sin x} = y$$

$$\sin x = 2y - y\sin x$$

$$y\sin x + \sin x = 2y$$

$$(\sin x)(y + 1) = 2y$$

$$\sin x = \frac{2y}{y + 1}$$

$$x = \arcsin\left(\frac{2y}{y + 1}\right).$$

Thus, $j^{-1}(x) = \arcsin\left(\dfrac{2x}{x + 1}\right)$.

33. (a) This is an increasing function, because as x increases, $f(x)$ increases, and as $f(x)$ increases, $f(f(x))$ increases.

(b) This is a decreasing function, because as x increases, $g(x)$ decreases, and as $g(x)$ decreases, $f(g(x))$ decreases.

(c) This is an increasing function, because as x increases, $g(x)$ decreases, and as $g(x)$ decreases, $g(g(x))$ increases.

(d) This is a decreasing function, because as x increases, $f(x)$ increases, and as $f(x)$ increases, $g(x)$ decreases.

(e) We can't tell. For example, suppose $f(x) = 2x$ and $g(x) = -x$. Then $f(x) + g(x) = x$, which is increasing. But if $f(x) = 2x$ and $g(x) = -3x$, then $f(x) + g(x) = -x$, which is decreasing.

(f) This is an increasing function, because if $g(x)$ is a decreasing function, then $-g(x)$ will be an increasing function. Since $f(x) - g(x) = f(x) + [-g(x)]$, $f(x) - g(x)$ can be written as the sum of two increasing functions, and is thus increasing.

CHAPTER SEVEN

Solutions for Section 7.1

1. Larger powers of x give smaller values for $0 < x < 1$.
   A - (iii)
   B - (ii)
   C - (iv)
   D - (i)

5. (a) As $x \longrightarrow 0$ from the right, $x^{-3} \longrightarrow +\infty$, and $x^{1/3} \longrightarrow 0$.
   (b) As $x \longrightarrow \infty$, $x^{-3} \longrightarrow 0$, and $x^{1/3} \longrightarrow \infty$.

9. (a) We are given that if d is the radius of the earth,

$$180 = \frac{k}{d^2},$$

so

$$d^2 = \frac{k}{180}.$$

On a planet whose radius is three times the radius of the earth, the person's weight is

$$w = \frac{k}{(3d)^2} = \frac{k}{9d^2}.$$

Therefore,

$$w = \frac{k}{9(\frac{k}{180})} = \frac{180}{9} = 20 \text{ lbs.}$$

If the radius is one-third of the earth's,

$$w = \frac{k}{(\frac{1}{3}d)^2} = \frac{k}{\frac{1}{9}d^2} = \frac{9k}{d^2} = \frac{9k}{\frac{k}{180}}$$
$$= 9(180) = 1620 \text{ lbs.}$$

(b) Let x be the fraction of the earth's radius, d. Then

$$2000 = \frac{k}{(xd)^2} = \frac{k}{x^2 d^2}.$$

We can use the information from part (a) to substitute

$$d^2 = \frac{k}{180}$$

so that

$$2000 = \frac{k}{x^2(\frac{k}{180})} = \frac{180}{x^2}.$$

Thus

$$x^2 = \frac{180}{2000} = 0.09$$

and

$$x = 0.3 \qquad \text{(discarding negative } x\text{)}.$$

The radius is $\frac{3}{10}$ of the earth's radius.

13. (a) The function f is the transformation of $y = \dfrac{1}{x}$, so $p = 1$. The graph of $y = \dfrac{1}{x}$ has been shifted three units to the right and four units up. To find the y-intercept, we need to evaluate $f(0)$:

$$f(0) = \frac{1}{-3} + 4 = \frac{11}{3}.$$

To find the x-intercepts, we need to solve $f(x) = 0$ for x.

$$\text{Thus,} \qquad 0 = \frac{1}{x-3} + 4,$$

$$-4 = \frac{1}{x-3},$$

$$-4(x-3) = 1,$$

$$-4x + 12 = 1,$$

$$-4x = -11,$$

$$\text{so} \qquad x = \frac{11}{4} \quad \text{is the only } x\text{-intercept.}$$

The graph of f is shown in Figure 7.1.

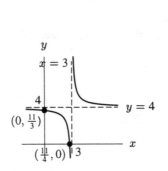

Figure 7.1

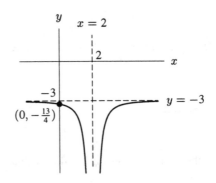

Figure 7.2

(b) The function g is a transformation of $y = \dfrac{1}{x^2}$, so $p = 2$. The graph of $y = \frac{1}{x^2}$ has been shifted 2 units to the right, flipped over the x-axis and shifted 3 units down. To find the y-intercept, we need to evaluate $g(0)$:

$$g(0) = -\frac{1}{4} - 3 = -\frac{13}{4}.$$

Note that this function has no x-intercepts.

The graph of g is shown in Figure 7.2.

(c) First, we can simplify the formula for $h(x)$:

$$h(x) = \frac{1}{x-1} + \frac{2}{1-x} + 2$$

$$= \frac{1}{x-1} - \frac{2}{x-1} + 2 = -\frac{1}{x-1} + 2.$$

Thus h is a transformation of $y = \dfrac{1}{x}$ with $p = 1$. The graph of $y = \frac{1}{x}$ has been shifted one unit to the right, flipped over the x-axis and shifted up 2 units. To find the y-intercept, we need to evaluate $h(0)$:

$$h(0) = -\frac{1}{-1} + 2 = 3.$$

To find the x-intercepts, we need to solve $h(x) = 0$ for x:

$$0 = -\frac{1}{x-1} + 2$$

$$-2 = -\frac{1}{x-1}$$

$$-2x + 2 = -1$$

$$-2x = -3,$$

$$\text{so,} \quad x = \frac{3}{2} \quad \text{is the only } x\text{-intercept.}$$

The graph of h is shown in Figure 7.3.

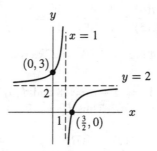

Figure 7.3

17. Note: $\frac{5}{7} > \frac{9}{16} > \frac{3}{8} > \frac{3}{11}$, so

$$A \longrightarrow kx^{5/7}, B \longrightarrow kx^{9/16}, C \longrightarrow kx^{3/8}, D \longrightarrow kx^{3/11}.$$

Solutions for Section 7.2

1. (a)

TABLE 7.1

x	$f(x)$	$g(x)$
-3	1/27	-27
-2	1/9	-8
-1	1/3	-1
0	1	0
1	3	1
2	9	8
3	27	27

(b) As $x \to -\infty$, $f(x) \to 0$. For f, large negative values of x result in small $f(x)$ values because a large negative power of 3 is very close to zero. For g, large negative values of x result in large negative values of $g(x)$, because the cube of a large negative number is a larger negative number. Therefore, as $x \to -\infty$, $g(x) \to -\infty$.

(c)   As $x \to \infty$, $f(x) \to \infty$ and $g(x) \to \infty$. For $f(x)$, large x-values result in large powers of 3; for $g(x)$, large x values yield the cubes of large x-values. f and g both climb *fast*, but f climbs faster than g (for $x > 3$).

5.   If $f(x) = mx^{1/3}$ goes through $(1, 2)$, then $m = 2$, so $f(x) = 2x^{1/3}$. Using $x = 8$ in $f(x) = 2x^{1/3}$ gives $t = 4$. If $g(x) = kx^{4/3}$ goes through $(8, 4)$, then $k = \frac{1}{4}$. Thus, $m = 2, t = 4$, and $k = \frac{1}{4}$.

9.   Solve for $g(x)$ by taking the ratio of (say) $g(4)$ to $g(3)$:

$$\frac{g(4)}{g(3)} = \frac{-32/3}{-9/2} = \frac{-32}{3} \cdot \frac{-2}{9} = \frac{64}{27}.$$

We know $g(4) = k \cdot 4^p$ and $g(3) = k \cdot 3^p$. Thus,

$$\frac{g(4)}{g(3)} = \frac{k \cdot 4^p}{k \cdot 3^p} = \frac{4^p}{3^p} = \left(\frac{4}{3}\right)^p = \frac{64}{27}.$$

Thus $p = 3$. To solve for k, note that $g(3) = k \cdot 3^3 = 27k$. Thus, $27k = g(3) = -\frac{9}{2}$. Thus, $k = -\frac{9}{54} = -\frac{1}{6}$. This gives $g(x) = -\frac{1}{6}x^3$.

13.   (a)   Let $f(x) = ax + b$. Then $f(1) = a + b = 18$ and $f(3) = 3a + b = 1458$. Solving simultaneous equations gives us $a = 720, b = -702$. Thus $f(x) = 720x - 702$.
     (b)   Let $f(x) = A \cdot B^x$, then

$$\frac{f(3)}{f(1)} = \frac{AB^3}{AB} = B^2 = \frac{1458}{18} = 81.$$

Thus,

$$B^2 = 81$$
$$B = 9 \qquad \text{(since } B \text{ must be positive)}$$

Using $f(1) = 18$ gives

$$A(9)^1 = 18$$
$$A = 2.$$

Therefore, if f is an exponential function, a formula for f would be

$$f(x) = 2(9)^x.$$

(c)   If f is a power function, let $f(x) = kx^p$, then

$$\frac{f(3)}{f(1)} = \frac{k(3)^p}{k(1)^p} = (3)^p$$

and

$$\frac{f(3)}{f(1)} = \frac{1458}{18} = 81.$$

Thus,

$$3^p = 81 \qquad \Rightarrow \qquad p = 4.$$

Solving for k, gives

$$18 = k(1^4) \qquad \Rightarrow \qquad k = 18.$$

Thus, a formula for f is

$$f(x) = 18x^4.$$

17. (a)

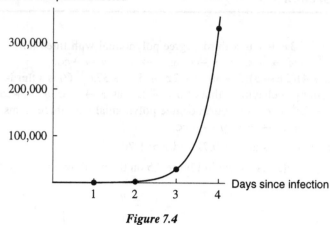

Number of computers infected

Figure 7.4

(b) The graph appears to be an exponential growth function.

(c) Note: $\frac{332,944}{29,311} \approx 11.359$, $\frac{29,311}{2580} \approx 11.361$ and $\frac{2580}{227} \approx 11.366$.

Thus, let $g(x) = a \cdot (11.36)^x$. To solve for a use any point, e.g. $g(1) = 227$. Then

$$227 = a(11.36)^1$$

so

$$a = \frac{227}{11.36} \approx 19.98.$$

Therefore, a model for g could be

$$g(x) = 19.98(11.36)^x.$$

(d) According to this model, approximately 20 computers were initially infected.

21. (a) Since $p = kd^{3/2}$ where k in this case is given by

$$k = \frac{365}{(93)^{3/2}}, \qquad \text{(in millions)}$$

so

$$p = \frac{365}{(93)^{3/2}} \cdot d^{3/2} = 365 \left(\frac{d}{93}\right)^{3/2}.$$

If p is twice the earth's period, $p = 2(365)$, so

$$2(365) = 365 \left(\frac{d}{93}\right)^{3/2}$$

$$2(93)^{3/2} = d^{3/2}$$

$$d^{3/2} \approx 1793.72\cdots$$

$$d \approx 147.6 \text{ million miles.}$$

(b) Yes, Mars orbits approximately 141 million miles from the sun.

Solutions for Section 7.3

1. (a) $y = 2x^3 - 3x + 7$ is a third-degree polynomial with three terms. Its long-range behavior is that of $y = 2x^3$: as $x \to -\infty, y \to -\infty$, as $x \to +\infty, y \to +\infty$.
    (b) $y = (x + 4)(2x - 3)(5 - x) = -2x^3 + 5x^2 + 37x - 60$ is a third-degree polynomial with four terms. Its long-range behavior is that of $y = -2x^3$: as $x \to -\infty, y \to +\infty$, as $x \to +\infty, y \to -\infty$.
    (c) $y = 1 - 2x^4 + x^3$ is a fourth degree polynomial with three terms. Its long-range behavior is that of $y = -2x^4$: as $x \to \pm\infty, y \to -\infty$.

5. There are two real zeros at $x \approx 0.72$ and $x \approx 1.70$.

9. The graph of $y = g(x)$ is shown in Figure 7.5 on the window $-5 \le x \le 5$ by $-20 \le y \le 10$.

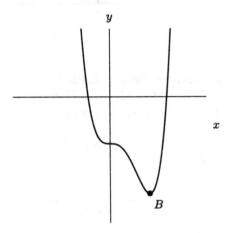

Figure 7.5

The minimum value of g occurs at point B as shown in the figure. Using either a table feature or trace on a graphing calculator, we approximate the minimum value of g to be -16.54 (to two decimal places).

13. (a) The graph of the function on the suggested window is shown in Figure 7.6.

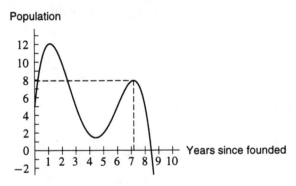

Figure 7.6

At $x = 0$ (when Smallsville was founded), the population was 5 hundred people.
    (b) The x-intercept for $x > 0$ will show when the population was zero. This occurs at $x \approx 8.44$. Thus, Smallsville became a ghost town in May of 1908.

(c) There are two peaks on the graph on $0 \le x \le 10$, but the first occurs before $x = 5$ (i.e.,before 1905). The second peak occurs at $x \approx 7.18$. The population at that point is ≈ 7.9 hundred. So the maximum population was ≈ 790 in February of 1907.

17. Yes. For the sake of illustration, suppose $f(x) = x^2 + x + 1$, a second-degree polynomial. Then

$$f(g(x)) = (g(x))^2 + g(x) + 1$$
$$= g(x) \cdot g(x) + g(x) + 1.$$

Since $f(g(x))$ is formed from products and sums involving the polynomial g, the composition $f(g(x))$ is also a polynomial. In general, $f(g(x))$ will be a sum of powers of $g(x)$, and thus $f(g(x))$ will be formed from sums and products involving the polynomial $g(x)$. A similar situation holds for $g(f(x))$, which will be formed from sums and products involving the polynomial $f(x)$. Thus, either expression will yield a polynomial.

Solutions for Section 7.4

1. The graph in Figure 7.51 in the text represents a polynomial of even degree, degree at least 4. Zeros are shown at $x = -2$, $x = -1$, $x = 2$, and $x = 3$. The leading coefficient must be negative. Thus, of the choices in Table 7.16, only C and E are possibilities. When $x = 0$, function C gives

$$y = -\frac{1}{2}(2)(1)(-2)(-3) = -\frac{1}{2}(12) = -6,$$

and function E gives

$$y = -(2)(1)(-2)(-3) = -12.$$

Since the y-intercept appears to be $(0, -6)$ rather than $(0, -12)$, function C best fits the polynomial shown.

5. (a) The graph appears to have x intercepts at $x = -\frac{1}{2}$, 3, 4, so let

$$f(x) = k(x + \frac{1}{2})(x - 3)(x - 4).$$

The y intercept is at $(0, 3)$, so

$$3 = f(0) = k(\frac{1}{2})(-3)(-4),$$
$$\text{which gives} \quad 3 = 6k,$$
$$\text{or} \quad k = \frac{1}{2}.$$

Therefore, $f(x) = \frac{1}{2}(x + \frac{1}{2})(x - 3)(x - 4)$ is a possible formula for f.

(b) The x intercepts are at $x = -4, -2, 2$, so let $f(x) = k(x + 4)(x + 2)(x - 2)$. The y intercept is at $(0, 24)$, so

$$24 = f(0) = k(4)(2)(-2)$$
$$24 = k(-16)$$
$$k = \frac{24}{-16} = \frac{3}{-2}.$$

Therefore, $f(x) = -\frac{3}{2}(x + 4)(x + 2)(x - 2)$ is a possible formula.

9. The function has a common factor of $4x$ which gives

$$f(x) = 4x(2x^2 - x - 15),$$

and the quadratic factor reduces further giving

$$f(x) = 4x(2x + 5)(x - 3).$$

Thus, the zeros of f are $x = 0$, $x = \frac{-5}{2}$, and $x = 3$.

13. (a) $V(x) = x(6 - 2x)(8 - 2x)$
     (b) Values of x for which $V(x)$ makes sense are $0 < x < 3$, since if $x < 0$ or $x > 3$ the volume is negative.
     (c)

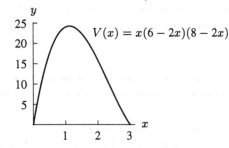

     (d) Using a graphing calculator, we find the peak between $x = 0$ and $x = 3$ to occur at $x \approx 1.13$. The maximum volume is ≈ 24.26 in^3.

17. (a) Factoring $y = x^2 + 5x + 6$ gives $y = (x + 2)(x + 3)$. Thus $y = 0$ for $x = -2$ or $x = -3$.
     (b) Note that $y = x^4 + 6x^2 + 9 = (x^2 + 3)^2$. This implies that $y = 0$ if $x^2 = -3$, but $x^2 = -3$ has no real solutions. Thus, there are no real zeros.
     (c) $y = 4x^2 - 1 = (2x - 1)(2x + 1)$, which implies that $y = 0$ for $x = \pm\frac{1}{2}$.
     (d) $y = 4x^2 + 1 = 0$ implies that $x^2 = -\frac{1}{4}$, which has no solutions. There are no real zeros.
     (e) By using the quadratic formula, we find that $y = 0$ if

$$x = \frac{3 \pm \sqrt{9 - 4(2)(-3)}}{4} = \frac{3 \pm \sqrt{33}}{4}.$$

     (f) This problem cannot be solved algebraically. Note that we cannot use the quadratic formula, as this is a 5th degree polynomial and not a 2nd degree polynomial. A graph of the function is shown in Figure 7.7 for $-1 \le x \le 1$, $-10 \le y \le 10$. From the graph, we approximate the zero to be at $x \approx -0.143$.

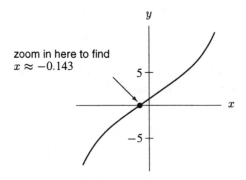

Figure 7.7

21. (a) We could let $f(x) = k(x + 2)(x - 3)(x - 5)$ to satisfy the given zeros. For a y-intercept of 4, $f(0) = 4 = k(0 + 2)(0 - 3)(0 - 5) = 30k$. Thus $30k = 4$, so $k = \frac{2}{15}$. One possibility is

$$f(x) = \frac{2}{15}(x + 2)(x - 3)(x - 5).$$

(b) One possibility is that f looks like the function in Figure 7.8.

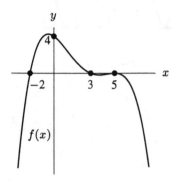

Figure 7.8

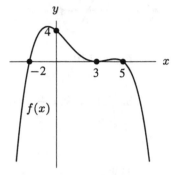

Figure 7.9

Then

$$f(x) = k(x + 2)(x - 3)(x - 5)^2$$

and

$$f(0) = k(2)(-3)(-5)^2.$$

Thus,

$$-150k = 4$$

which gives us

$$k = -\frac{2}{75}.$$

Thus

$$f(x) = -\frac{2}{75}(x + 2)(x - 3)(x - 5)^2.$$

Another possibility is that f has a double-zero at $x = 3$ instead of at $x = 5$. In this case f looks like the function in Figure 7.9. This would give the formula

$$f(x) = -\frac{2}{45}(x + 2)(x - 3)^2(x - 5).$$

Note that if f had a double zero at $x = -2$, there must be another zero for $-2 < x < 0$ in order for f to satisfy $f(0) = 4$ and $y \to -\infty$ as $x \to \pm\infty$.

(c) One possibility is that f looks like the graph in Figure 7.10.

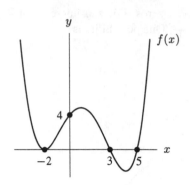

Figure 7.10

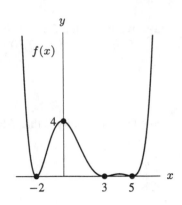

Figure 7.11

So $f(x) = k(x+2)^2(x-3)(x-5)$, which gives us

$$k = \frac{1}{15}.$$

Thus,

$$f(x) = \frac{1}{15}(x+2)^2(x-3)(x-5).$$

It is also possible that f has 3 double-zeros at $x = -2$, $x = 3$ and $x = 5$. This leads to a 6th degree polynomial which looks like Figure 7.11. This would give the formula

$$f(x) = \frac{1}{225}(x+2)^2(x-3)^2(x-5)^2.$$

Solutions for Section 7.5

1. (a) (i) $C(1) = 5050$ means the cost to make 1 unit is $5050.
   (ii) $C(100) = 10,000$ means the cost to make 100 units is $10,000.
   (iii) $C(1000) = 55,000$ means the cost to make 1000 units is $55,000.
   (iv) $C(10000) = 505,000$ means the cost to make 10,000 units is $505,000.
   (b) (i) $A(1) = \frac{C(1)}{1} = 5050$ means that it costs $5050/unit to make 1 unit.
   (ii) $A(100) = \frac{C(100)}{100} = 100$ means that it costs $100/unit to make 100 units.
   (iii) $A(1000) = \frac{C(1000)}{1000} = 55$ means that it costs $55/unit to make 1000 units.
   (iv) $A(10000) = \frac{C(10000)}{10000} = 50.5$ means that it costs $50.50/unit to make 10,000 units.
   (c) As the number of units increases, the average cost per unit gets closer to $50/unit, which is the unit (or marginal) cost. This makes sense because the fixed or initial $5000 expenditure becomes increasingly insignificant as it is averaged over a large number of units.

5. For the function f, $f(x) \to 1$ as $x \to \pm\infty$ since for large values of x, $f(x) \approx \frac{x^2}{x^2} = 1$.

   The function $g(x) \approx \frac{x^3}{x^2} = x$ for large values of x. Thus, as $x \to \pm\infty$, $g(x)$ approaches the line $y = x$.

   The function h will behave like $y = \frac{x}{x^2} = \frac{1}{x}$ for large values of x. Thus, $h(x) \to 0$ as $x \to \pm\infty$.

9. (a) Figure 7.66 in the text appears to be $y = \frac{1}{x^2}$ shifted 3 units to the right and flipped across the x-axis. Thus,

$$y = -\frac{1}{(x-3)^2}$$

is a formula.

(b) The equation $y = -\frac{1}{(x-3)^2}$ can be written as

$$y = \frac{-1}{x^2 - 6x + 9}.$$

(c) Since y can not equal zero if $y = \frac{-1}{x^2-6x+9}$, Figure 7.66 has no x-intercept. The y-intercept occurs when $x = 0$, so $y = \frac{-1}{(-3)^2} = -\frac{1}{9}$. The y-intercept is at $(0, -\frac{1}{9})$.

13. (a) Table 7.19 in the problem shows a translation of $y = \frac{1}{x^2}$. Table 7.19 shows symmetry about the vertical asymptote $x = 3$. The fact that the function values have the same sign on both sides of the vertical asymptotes indicates a transformation of $y = \frac{1}{x^2}$ rather than $y = \frac{1}{x}$.

(b) In order to shift the vertical asymptote from $x = 0$ to $x = 3$ for Table 7.19, we try

$$y = \frac{1}{(x-3)^2}.$$

checking the x-values from the table in this formula gives y-values that are each 1 less than the y-values of the table. Therefore, we try

$$y = \frac{1}{(x-3)^2} + 1.$$

This formula works. To express the formula as a ratio of polynomials, we take

$$y = \frac{1}{(x-3)^2} + \frac{1(x-3)^2}{(x-3)^2},$$

so

$$y = \frac{x^2 - 6x + 10}{x^2 - 6x + 9}.$$

Solutions for Section 7.6

1. (a) The zero of this function is at $x = -3$.
   It has a vertical asymptote at $x = -5$.
   Its long-range behavior is: $y \to 1$ as $x \to \pm\infty$.

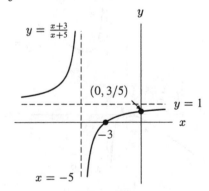

$$y = \frac{x+3}{x+5}$$

Figure 7.12

(b) The zero of this function is at $x = -3$.
It has a vertical asymptote at $x = -5$.
Its long-range behavior is: $y \to 0$ as $x \to \pm\infty$.

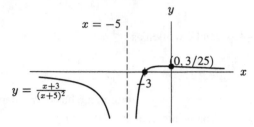

Figure 7.13

(c) The zero of this function is at $x = 4$.
It has vertical asymptotes at $x = \pm 3$.
Its long-range behavior is: $y \to 0$ as $x \to \pm\infty$.

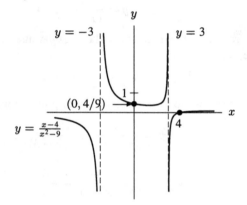

Figure 7.14

(d) The zeros of this function are at $x = \pm 2$.
It has a vertical asymptote at $x = 9$.
Its long-range behavior is that it looks like the line $y = x$.

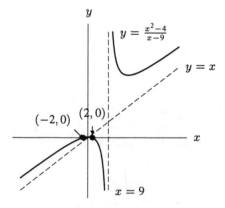

Figure 7.15

5. (a) The graph of $y = \frac{1}{f(x)}$ will have vertical asymptotes at $x = 0$ and $x = 2$. As $x \to 0$ from the left, $\frac{1}{f(x)} \to -\infty$, and as $x \to 0$ from the right, $\frac{1}{f(x)} \to +\infty$. The reciprocal of 1 is 1, so $\frac{1}{f(x)}$ will also go through the point (1,1). As $x \to 2$ from the left, $\frac{1}{f(x)} \to +\infty$, and as $x \to 2$ from the right, $\frac{1}{f(x)} \to -\infty$. As $x \to \pm\infty$, $\frac{1}{f(x)} \to 0$ and is negative.

The graph of $y = \frac{1}{f(x)}$ is shown in Figure 7.16.

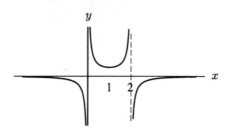

Figure 7.16

(b) A formula for f is of the form

$$f(x) = k(x - 0)(x - 2) \quad \text{and} \quad f(1) = 1.$$

Thus, $1 = k(1)(-1)$, so $k = -1$. Thus

$$f(x) = -x(x - 2).$$

The reciprocal $\frac{1}{f(x)} = -\frac{1}{x(x-2)}$ is graphed as shown in Figure 7.16.

9. (a) We see that the graph has a double zero at $x = 0$, and zeros at $x = -3$ and $x = 2$. So let

$$f(x) = kx^2(x + 3)(x - 2).$$

Since $f(-2) = -1$, we have

$$f(-2) = k(-2)^2(-2 + 3)(-2 - 2) = -16k$$

which gives

$$-16k = -1,$$
$$\text{so} \quad k = \frac{1}{16}.$$

Thus a possible formula is

$$f(x) = \frac{1}{16}x^2(x + 3)(x - 2).$$

(b) • Since the graph has an asymptote at $x = 2$, let the denominator be $(x - 2)$.
    • Since the graph has a zero at $x = -1$, let the numerator be $(x + 1)$.
    • Since the long–range behavior tends toward -1 as $x \to \pm\infty$, the ratio of the leading terms should be -1.

So a possible formula is $y = f(x) = -\left(\dfrac{x + 1}{x - 2}\right)$. You can check that the y–intercept is $y = \frac{1}{2}$, as it should be.

(c) • Since the graph has asymptotes at $x = -1$ and $x = 2$, let the denominator be $(x - 2)(x + 1)$.
   • Since the graph has zeros at $x = -2$ and $x = 3$, let the numerator be $(x + 2)(x - 3)$.
   • Since the long–range behavior tends toward 1 as $x \to \pm\infty$, the ratio of the leading terms should be 1.

   So a possible formula is $y = f(x) = \dfrac{(x + 2)(x - 3)}{(x - 2)(x + 1)}$. You can check that the y-intercept is $y = 3$, as it should be.

13. (a) $k(x) = \frac{1}{g(x)}$, or $k(x) = \frac{f(x)}{g(x)^2}$
    (b) $m(x) = \frac{f(x)}{g(x)}$
    (c) $n(x) = \frac{1}{f(x)}$
    (d) $p(x) = \frac{g(x)}{f(x)}$
    (e) $q(x) = \frac{f(x)}{h(x)}$

17. (a) If $A < C < B < D$, we would have a graph with a zero at $x = A$ followed by vertical asymptote of $x = C$, another zero at $x = B$ and finally an asymptote of $x = D$. None of the graphs fit this pattern.
    (b) The graph of $y = g(x)$ shows two distinct zeros followed by two vertical asymptotes. Thus, g fits the pattern of $A < B < C < D$.
    (c) If $A < C < D < B$, we would need a graph with a zero, followed by two distinct asymptotes, followed by a zero. None of the graphs fits this pattern.
    (d) The graph of $y = f(x)$ has an asymptote followed by a repeated zero and then another asymptote. Thus, f fits $C < A$, $A = B$, $B < D$.
    (e) The pattern $A < C$, $C = D$, $D < B$ indicates a zero followed by a single asymptote followed by a zero. Thus, $y = j(x)$ fits this pattern.
    Note that there is not a statement to match $y = h(x)$. A possible pattern for h would be $C < A < B < D$, which $A = -B$.

Solutions for Chapter 7 Review

1. (a) The y coordinates in table (a) fluctuate between 3 and -1. A trigonometric function would be a good model for this type of behavior. Since at $x = 0$ the graph would be at a peak, a cosine seems appropriate. The amplitude will be 2 and the mid-line will be $y = 1$. The period is 2. Thus,

$$y = 2\cos(\pi x) + 1$$

is one possible choice. [Note: This answer is not the only possible choice.]
   (b) Only positive values of x are shown. For $0 < x < 1$ the x-values have negative y coordinates. The function passes through $(1,0)$ and is increasing—albeit slowly. All of these features fit a logarithmic model. Then we would have a formula of the form $y = \log_b x$. Since $y = 1$ when $x = 3$, this gives $b = 3$. A check of $y = \log_3 x$ fits the data of table (b).
   (c) Table (c) shows 3 values of x for which $y = 0$. The function does not appear periodic, so a polynomial may be the best choice. The zeros at $x = -2, x = 1$ and $x = 2$ suggest a cubic of the form

$$y = k(x + 2)(x - 1)(x - 2).$$

Since $y = 8$ when $x = 0$,

$$8 = k(2)(-1)(-2)$$
$$k = 2.$$

Note that

$$y = 2(x + 2)(x - 1)(x - 2)$$

fits the data of table (c) exactly.

(d) The symmetry of the y coordinates leads us to consider an odd function. The data is clearly not linear, nor does it indicate periodic behavior. A rough sketch suggests a cubic power function. Note the function has not been shifted horizontally or vertically, so a good guess might be $y = kx^3$. Using the point $(1,3)$ gives $k = 3$. Try

$$y = 3x^3.$$

[Yes—works great!]

(e) The data points indicate an increasing function—certainly not linear. In fact, the function values are increasing by greater and greater amounts as x increases. Try an exponential function. A look at the ratios of successive y-values shows a constant ratio of 5. The y-intercept of 0.5 indicates that

$$y = 0.5(5)^x$$

may be appropriate.— It fits beautifully!

(f) Well, if elimination works, there must be a linear function for the last table. We wouldn't need to have guessed, however. Note that $\frac{\Delta y}{\Delta x}$ is consistently $-\frac{5}{1}$. Thus, the slope is -5 and the model should be of the form

$$y = b - 5x.$$

Using any data point (e.g., $(1, 8)$) we solve for b:

$$8 = b - 5$$
$$b = 13.$$

Thus,

$$y = 13 - 5x.$$

Check it out!

5. (a) If q is a second degree polynomial with zeros at $x = a$ and $x = b$, we know q has x-intercepts at $x = a$ and $x = b$. Since q is a second degree polynomial, we also know $y = q(x)$ is graphically a parabola. However, we basically know nothing else about the graph of q. Without further information, we cannot tell whether the parabola is opening up or down or what the y-coordinate of the vertex is (although the vertex will be on the line $x = \frac{|b-a|}{2}$, that is, half way between the x-intercepts).

(b) The factored form of q must contain factors $(x - a)$ and $(x - b)$, however, $q(x) = (x - a)(x - b)$ is a *particular* parabola which assumes information not given. To account for the general case, we let $q(x) = k(x - a)(x - b)$. With an additional point on the parabola, we could solve for k.

9. (a) If f is even, then for all x

$$f(x) = f(-x).$$

Since $f(x) = ax^2 + bx + c$, this implies

$$ax^2 + bx + c = a(-x)^2 + b(-x) + c$$
$$= ax^2 - bx + c$$

We can cancel the ax^2 term and the constant term c from both sides of this equation, giving

$$bx = -bx$$
$$bx + bx = 0$$
$$2bx = 0.$$

Since x is not necessarily zero, we conclude that b must equal zero, so that if f is even,

$$f(x) = ax^2 + c.$$

(b) If g is odd, then for all x

$$g(-x) = -g(x).$$

Since $g(x) = ax^3 + bx^2 + cx + d$, this implies

$$
\begin{aligned}
ax^3 + bx^2 + cx + d &= -(a(-x)^3 + b(-x)^2 + c(-x) + d) \\
&= -(-ax^3 + bx^2 - cx + d) \\
&= ax^3 - bx^2 + cx - d.
\end{aligned}
$$

The odd-powered terms cancel, leaving

$$
\begin{aligned}
bx^2 + d &= -bx^2 - d \\
2bx^2 + 2d &= 0 \\
bx^2 + d &= 0.
\end{aligned}
$$

Since this must hold true for any value of x, we know that both b and d must equal zero. Therefore,

$$g(x) = ax^3 + cx.$$

13.

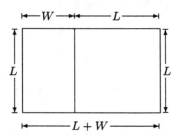

Figure 7.17: These two rectangles have the same proportions

Since the two rectangles have the same proportions, we know

$$\frac{L}{W} = \frac{L+W}{L}.$$

Writing $\frac{L+W}{L} = \frac{L}{L} + \frac{W}{L} = 1 + \frac{W}{L} = 1 + \frac{1}{\frac{L}{W}}$, we have

$$\frac{L}{W} = 1 + \frac{1}{\left(\frac{L}{W}\right)}.$$

We know that the Golden Ratio, ϕ, is given by $\frac{L}{W}$. Thus, we have

$$\phi = 1 + \frac{1}{\phi}.$$

Multiplying both sides by ϕ, we have

$$\phi^2 = \phi + 1,$$
$$\phi^2 - \phi - 1 = 0.$$

Using the quadratic formula, we have

$$\phi = \frac{1 \pm \sqrt{1 - 4(1)(-1)}}{2} = \frac{1 \pm \sqrt{5}}{2}.$$

Since ϕ is a ratio of two lengths, it must be positive. Thus,

$$\phi = \frac{1 + \sqrt{5}}{2}.$$

17. (a) If $y = f(x)$, then $x = f^{-1}(y)$. Solving $y = f(x)$ for x, we have

$$y = \frac{x}{x + 5}$$
$$y(x + 5) = x$$
$$yx + 5y = x$$
$$yx - x = -5y$$
$$x(y - 1) = -5y$$
$$x = \frac{-5y}{y - 1} = \frac{5y}{1 - y}.$$

Thus, $f^{-1}(x) = \frac{5x}{1-x}$.

(b) $f^{-1}(0.2) = \frac{5(0.2)}{(1-0.2)} = \frac{1}{0.8} = 1.25$. This means that 1.25 gallons of alcohol must be added to give an alcohol concentration of .20 or 20%.

(c) $f^{-1}(x) = 0$ means that $\frac{5x}{1-x} = 0$ which means that $x = 0$. This means that 0 gallons of alcohol must be added to give a concentration of 0%.

(d) The horizontal asymptote of $f^{-1}(x)$ is $y = -5$. Since x is a concentration of alcohol, $0 \le x \le 1$. Thus, the regions of the graph for which $x < 0$ and $x > 1$ have no physical significance. Consequently, since $f^{-1}(x)$ approaches its asymptote only as $x \to \pm\infty$, its horizontal asymptote has no physical significance.

21. (a)

$$p(x) = k(x + 3)(x - 2)(x - 5)(x - 6)^2$$
$$7 = p(0) = k(3)(-2)(-5)(-6)^2$$
$$= k(1080)$$
$$k = \frac{7}{1080}$$
$$p(x) = \frac{7}{1080}(x + 3)(x - 2)(x - 5)(x - 6)^2$$

(b)

$$f(x) = \frac{p(x)}{q(x)} = \frac{(x + 3)(x - 2)}{(x + 5)(x - 7)}$$

(c)

$$f(x) = \frac{p(x)}{q(x)} = \frac{-3(x - 2)(x - 3)}{(x - 5)^2}$$

We need the factor of -3 in the numerator and the exponent of 2 in the denominator, because we have a horizontal asymptote of $y = -3$. The ratio of highest term of $p(x)$ to highest term of $q(x)$ will be $\frac{-3x^2}{x^2} = -3$.

25. (a) $f(x) = -\dfrac{1}{(x-5)^2} - 1$ has vertical asymptote at $x = 5$, no x intercept, horizontal asymptote $y = -1$:
     (iii)

     (b) vertical asymptotes at $x = -1, 3$, x intercept at 2, horizontal asymptote $y = 0$: (i)

     (c) vertical asymptotes at $x = 1$, x intercept at $x = -2$, horizontal asymptote $y = 2$: (ii)

     (d) $f(x) = \dfrac{x - 3 + x + 1}{(x+1)(x-3)} = \dfrac{2x - 2}{(x+1)(x-3)}$ has vertical asymptotes at $x = -1, 3$, x intercept at 1, horizontal asymptote at $y = 0$: (iv)

     (e) $f(x) = \dfrac{(1+x)(1-x)}{x-2}$ has vertical asymptote at $x = 2$, two x intercepts at ± 1: (vi)

     (f) vertical asymptote at $x = -1$, x intercept at $x = \frac{1}{4}$, horizontal asymptote at $y = -2$: (v)

29. (a) Let $y = f^{-1}(x)$. Then $x = f^{-1}(y)$. Solving $y = f(x) = \dfrac{3x}{x-1}$ for x, we have

$$\frac{3x}{x-1} = y$$
$$3x = xy - y$$
$$3x - xy = -y$$
$$x(3 - y) = -y$$
$$x = \frac{-y}{3 - y} = \frac{y}{y - 3}$$

Thus, $f^{-1}(y) = \dfrac{y}{y-3}$, which means that $f^{-1}(x) = \dfrac{x}{x-3}$.

     (b) Let $y = g(x)$. Then, $x = g^{-1}(y)$. Solving $y = g(x) = (x-3)^2 + 4$ for x, we have

$$(x - 3)^2 + 4 = y$$
$$(x - 3)^2 = y - 4$$
$$x - 3 = \pm\sqrt{y - 4}.$$

Since $x \geq 3$, $x - 3$ must not be negative, which gives

$$x - 3 = \sqrt{y - 4}$$
$$x = 3 + \sqrt{y - 4}.$$

Thus, $g^{-1}(y) = 3 + \sqrt{y - 4}$, which means that $g^{-1}(x) = 3 + \sqrt{x - 4}$.

     (c) Let $y = h(x)$. Then $x = h^{-1}(y)$. Solving $y = h(x) = 1 - \dfrac{2}{x-3}$ for x, we have

$$1 - \frac{2}{x - 3} = y$$
$$x - 3 - 2 = y(x - 3)$$
$$x - 5 = xy - 3y$$
$$x - xy = 5 - 3y$$
$$x(1 - y) = 5 - 3y$$
$$x = \frac{5 - 3y}{1 - y} = \frac{3y - 5}{y - 1}.$$

Thus, $h^{-1}(y) = \dfrac{3y - 5}{y - 1}$, which means that $h^{-1}(x) = \dfrac{3x - 5}{x - 1}$.

(d) Let $y = j(x)$. Then, $x = j^{-1}(y)$. Solving $y = j(x) = \dfrac{2x^3 + 1}{3x^3 - 1}$ for x, we have

$$\frac{2x^3 + 1}{3x^3 - 1} = y$$
$$2x^3 + 1 = y(3x^3 - 1) = 3x^3y - y$$
$$2x^3 - 3x^3y = -1 - y$$
$$x^3(2 - 3y) = -1 - y$$
$$x^3 = \frac{-1 - y}{2 - 3y} = \frac{y + 1}{3y - 2}$$
$$x = \sqrt[3]{\frac{y + 1}{3y - 2}}.$$

Thus, $j^{-1}(y) = \sqrt[3]{\dfrac{y + 1}{3y - 2}}$, which means that $j^{-1}(x) = \sqrt[3]{\dfrac{x + 1}{3x - 2}}$.

(e) Let $y = k(x)$. then $x = k^{-1}(y)$. Solving $y = k(x) = (\sqrt{x} + 1)^3$ for x, we have

$$(\sqrt{x} + 1)^3 = y$$
$$\sqrt{x} + 1 = \sqrt[3]{y}$$
$$\sqrt{x} = \sqrt[3]{y} - 1$$
$$x = (\sqrt[3]{y} - 1)^2.$$

Thus, $k^{-1}(y) = (\sqrt[3]{y} - 1)^2$, which means that $k^{-1}(x) = (\sqrt[3]{x} - 1)^2$.

(f) Let $y = l(x)$. Then $x = l^{-1}(y)$. Solving $y = l(x) = x^2 + 4x$ for x, we have

$$x^2 + 4x = y.$$

We can solve this equation by completing the square.

$$x^2 + 4x + 4 = y + 4$$
$$(x + 2)^2 = y + 4$$
$$x + 2 = \pm\sqrt{y + 4}$$

But since $x \geq -2$, we know that $x + 2$ cannot be negative. Thus,

$$x + 2 = \sqrt{y + 4}$$
$$x = -2 + \sqrt{y + 4}$$

Thus $l^{-1}(y) = \sqrt{y + 4} - 2$, which means that $l^{-1}(x) = \sqrt{x + 4} - 2$.

33. (a)

TABLE 7.2

x	$p(x)$	$\Delta p(x)$
0	0	
1	1	1
2	8	7
3	27	19
4	64	37

(b)   We can write $\Delta p(x)$ as follows

$$\begin{aligned}
\Delta p(x) &= p(x) - p(x-1) \\
&= x^3 - (x-1)^3 \\
&= x^3 - (x^3 - 3x^2 + 3x - 1) \\
&= 3x^2 - 3x + 1
\end{aligned}$$

Plugging in some values we see that this function and the table agree.

37.   (a)   Since x is the proportion of the population that is infected, the number infected is xP. The proportion of the population that is not infected is $1 - x$. Thus, the number not infected is $(1-x)P$.

(b)   We are given that the sensitivity of the test is 95%, which means that 19 out of 20 infected people will test positive. As there are xP infected people, the number of true positives is $19xP/20$. Because the specificity of the test is 90%, 10% or 1 out of 10 of those not infected test positive. As there are $(1-x)P$ people who are not infected, the number of false positives is $(1-x)P/10$.

(c)   The predictive value, $V(x)$, is the ratio of true positives to all positives. We know that the number of all positives is

$$\frac{19xP}{20} + \frac{(1-x)P}{10} = \frac{19xP + 2(1-x)P}{20} = \frac{17xP + 2P}{20}.$$

Thus, we have

$$V(x) = \frac{19xP/20}{(17xP + 2P)/20} = \frac{19xP}{17xP + 2P} = \frac{19x}{17x + 2}.$$

Note that the predictive value V depends only on the prevalence x, and not on the size of the population screened, P.

(d)

TABLE 7.3

x	$V(x)$
0.1%	0.94%
0.2%	1.87%
1.0%	8.76%
2.0%	16.24%
5.0%	33.33%
10.0%	51.35%
20.0%	70.37%
40.0%	86.36%
80.0%	97.44%

The predictive value is very sensitive to the prevalence when the prevalence is low. It seems that when the prevalence doubles, so does, roughly, the predictive value. For example, the table shows the prevalence doubling between 0.1% and 0.2%, and between 1% and 2%. In both cases, the predictive value doubles, too, from 0.94% to 1.87% and from 8.76% to 16.24%. However, this trend does not continue. The predictive value still increases as the prevalence increases, but the increase is not quite so rapid.

(e)   We need to find the value of x for which the predictive value is 5%, or 1/20. Thus, we have

$$\begin{aligned}
\frac{19x}{17x + 2} &= \frac{1}{20} \\
380x &= 17x + 2 \\
363x &= 2 \\
x &= \frac{2}{363} \approx 0.551\%.
\end{aligned}$$

41.   (a)

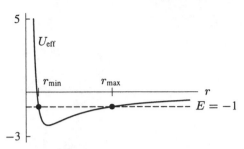

Figure 7.18: An energy diagram giving the
system's effective potential as well as total energy.

(b)   As you can see from Figure 7.18, the total energy E is greater than the effective potential U_{eff} for
      $r_{\min} < r < r_{\max}$. Using a computer or a graphing calculator to solve for $r_{\min}$ and $r_{\max}$, we have

$$r_{\min} \approx 0.382 \quad \text{and} \quad r_{\max} \approx 2.618.$$

Thus, the planet's perihelion is $r_{\min}$, or about 0.382.

(c)   The planet's aphelion is $r_{\max} \approx 2.618$.

45.   (a)   For the system whose total energy is E_2, the planet is fixed at a distance r_2 from its sun. This is because
      at any other distance r, the effective potential exceeds the total energy. Thus, this planet moves around
      its sun at a constant distance, and its orbit is a circle of radius r_2.

(b)   For the system whose total energy is E_1, the planet can come no closer to its sun than r_1, and can
      go no further away than r_3. Thus, this planet's orbit is an ellipse whose semimajor axis is of length
      $(r_1 + r_3)/2$. See Figure 7.19.

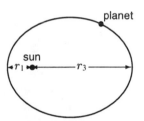

Figure 7.19

49.   (a)   The planet can be no closer to the sun than r_0, for otherwise its kinetic energy of radial motion would
      be negative. Thus, r_0 is the planet's perihelion.

(b)   This planet can move arbitrarily far away from the sun, since there is no distance r for which $E < U_{\text{eff}}$.
      Thus, it is not "bound" to the star, and is free to drift away.

CHAPTER EIGHT

Solutions for Section 8.1

1.

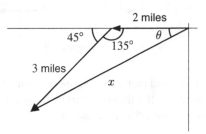

Figure 8.1

In Figure 8.1 we choose north as the positive y-axis, east as the positive x-axis and the house as the origin.

We know that the first vector is 2 units long, the second vector is 3 units and is at an angle of $45°$ from the first. Joining the tail of the first vector and the head of the second vector forms a triangle.

The length of the third side, x, can be found by applying the Law of Cosines:

$$x^2 = 2^2 + 3^2 - 2 \cdot 2 \cdot 3 \cdot \cos(135°) = 13 - 12 \left(-\frac{\sqrt{2}}{2} \right)$$

$$x^2 = 13 + 6\sqrt{2} = 21.4853$$

$$x = 4.63.$$

To obtain the angle θ, we apply the Law of Sines:

$$\frac{\sin \theta}{3} = \frac{\sin 135°}{x}$$

$$\sin \theta = 3 \cdot \frac{\sqrt{2}/2}{4.63} = 0.46.$$

So $\theta = 27.3°$.

Therefore, the person should walk $27.3°$ north of east for 4.63 miles to get back home. See Figure 8.1.

5.

$$\vec{p} = 2\vec{w}, \quad \vec{q} = -\vec{u}, \quad \vec{r} = \vec{w} + \vec{u} = \vec{u} + \vec{w},$$
$$\vec{s} = \vec{p} + \vec{q} = 2\vec{w} - \vec{u}, \quad \vec{t} = \vec{u} - \vec{w}$$

9. The wind velocity is a vector because it has both a magnitude (the speed of the wind) and a direction (the direction of the wind).

13.

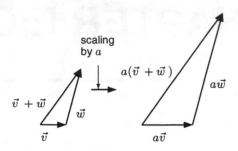

Figure 8.2

The effect of scaling the left-hand picture in Figure 8.2 is to stretch each vector by a factor of a (shown with $a > 1$). Since, after scaling up, the three vectors $a\vec{v}$, $a\vec{w}$, and $a(\vec{v} + \vec{w})$ form a similar triangle, we know that $a(\vec{v} + \vec{w})$ is the sum of the other two: that is

$$a(\vec{v} + \vec{w}) = a\vec{v} + a\vec{w}.$$

17. By Figure 8.3, the vectors $\vec{v} + (-1)\vec{w}$ and $\vec{v} - \vec{w}$ are equal.

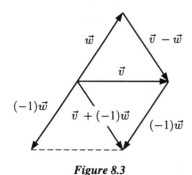

Figure 8.3

Solutions for Section 8.2

1. (a) $\vec{v} = 2\vec{i} + \vec{j}$
   (b) Since $\vec{w} = \vec{i} - \vec{j}$, we have $2\vec{w} = 2\vec{i} - 2\vec{j}$.
   (c) Since $\vec{v} = 2\vec{i} + \vec{j}$ and $\vec{w} = \vec{i} - \vec{j}$, we have $\vec{v} + \vec{w} = (2\vec{i} + \vec{j}) + (\vec{i} - \vec{j}) = 3\vec{i}$.
   (d) Since $\vec{w} = \vec{i} - \vec{j}$ and $\vec{v} = 2\vec{i} + \vec{j}$, we have $\vec{w} - \vec{v} = (\vec{i} - \vec{j}) - (2\vec{i} + \vec{j}) = -\vec{i} - 2\vec{j}$.
   (e) $\vec{PQ} = \vec{i} + \vec{j}$
   (f) Since P is at the point $(1, -2)$, the vector we want is $(2 - 1)\vec{i} + (0 - (-2))\vec{j} = \vec{i} + 2\vec{j}$.
   (g) The vector must be horizontal, so $\vec{i}$ will work.
   (h) The vector must be vertical, so $\vec{j}$ will work.

5. (a) The displacement from P to Q is given by

$$\vec{PQ} = (4\vec{i} + 6\vec{j}) - (\vec{i} + 2\vec{j}) = 3\vec{i} + 4\vec{j}.$$

Since
$$\|\overrightarrow{PQ}\| = \sqrt{3^2 + 4^2} = 5,$$

a unit vector $\vec{u}$ in the direction of $\overrightarrow{PQ}$ is given by
$$\vec{u} = \frac{1}{5}\overrightarrow{PQ} = \frac{1}{5}(3\vec{i} + 4\vec{j}) = \frac{3}{5}\vec{i} + \frac{4}{5}\vec{j}.$$

(b) A vector of length 10 pointing in the same direction is given by
$$10\vec{u} = 10(\frac{3}{5}\vec{i} + \frac{4}{5}\vec{j}) = 6\vec{i} + 8\vec{j}.$$

9. The vector $\vec{w}$ appears to consist of $9.2 - 6.3$ units to the left on the x-axis, and $4.5 - 0.7$ units down on the y-axis. Multiplying by 0.25 to convert to inches gives,
$$\vec{v} \approx -0.7\vec{i} - 1.0\vec{j}.$$

13.

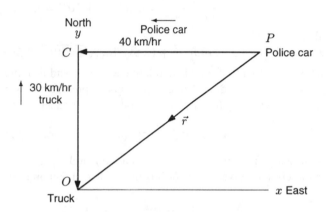

Figure 8.4

Since both vehicles reach the crossroad in exactly one hour, at the present the truck is at O in Figure 8.4; the police car is at P and the crossroads is at C. If $\vec{r}$ is the vector representing the line of sight of the truck with respect to the police car.
$$\vec{r} = -40\vec{i} - 30\vec{j}$$

17. (a) To be parallel, vectors must be scalar multiples. The $\vec{k}$ component of the first vector is 2 times the $\vec{k}$ component of the second vector. So the $\vec{i}$ components of the two vectors must be in a 2:1 ratio, and the same is true for the $\vec{j}$ components. Thus, $4 = 2a$ and $a = 2(a - 1)$. These equations have the solution $a = 2$, and for that value, the vectors are parallel.

(b) Perpendicular means a zero dot product. So $4a + a(a - 1) + 18 = 0$, or $a^2 + 3a + 18 = 0$. Since $b^2 - 4ac = 9 - 4 \cdot 1 \cdot 18 = -63 < 0$, there are no real solutions. This means the vectors are never perpendicular.

21. $\|\vec{v}\| = \sqrt{7.2^2 + (-1.5)^2 + 2.1^2} = \sqrt{58.5} \approx 7.6$

25.

$$\text{Displacement} = \text{Squirrel's coordinates} - \text{Cat's coordinates}$$
$$= (2 - 1)\vec{i} + (4 - 4)\vec{j} + (1 - 0)\vec{k} = \vec{i} + \vec{k}$$

Solutions for Section 8.3 ━━━━━━━━━━━━━━━━━━━━━━━━━━━━━━━━━━━━

1. (a)

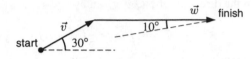

Figure 8.5

Since $\|\vec{v}\| = 5$,
$$\vec{v} = 5\cos 30°\vec{i} + 5\sin 30°\vec{j} = 4.33\vec{i} + 2.5\vec{j}.$$

For the second leg of his journey, $\vec{w} = x\vec{i}$.

(b) The vector from finish to start is $-(\vec{v} + \vec{w}) = (-4.33 - x)\vec{i} - 2.5\vec{j}$. This vector is at an angle of $10°$ south of west. So
$$\frac{-2.5}{-4.33 - x} = \tan(180° + 10°) = 0.176.$$

This means that $x = 9.87$.

(c) The distance home is $\|(-4.33 - 9.87)\vec{i} + (2.5)\vec{j}\| = \sqrt{14.20^2 + 2.5^2} = 14.42$.

5. Suppose $\vec{u}$ represents the velocity of the plane relative to the air and $\vec{w}$ represents the velocity of the wind. We can add these two vectors by adding their components. Suppose north is in the y-direction and east is the x-direction. The vector representing the airplane's velocity makes an angle of $45°$ with north; the components of $\vec{u}$ are
$$\vec{u} = 700\sin 45°\vec{i} + 700\cos 45°\vec{j} \approx 495\vec{i} + 495\vec{j}.$$

Since the wind is blowing from the west, $\vec{w} = 60\vec{i}$. By adding these we get a resultant vector $\vec{v} = 555\vec{i} + 495\vec{j}$. The direction relative to the north is the angle θ shown in Figure 8.6 given by
$$\theta = \tan^{-1}\frac{x}{y} = \tan^{-1}\frac{555}{495}$$
$$\approx 48.3°.$$

The magnitude of the velocity is
$$\|\vec{v}\| = \sqrt{495^2 + 555^2} = \sqrt{553{,}050}$$
$$= 744 \text{ km/hr.}$$

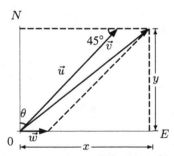

Figure 8.6: Note that θ is the angle between north and the vector $\vec{v}$

Solutions for Section 8.4

1. Since $3\vec{i} + \sqrt{3}\vec{j} = \sqrt{3}(\sqrt{3}\vec{i} + \vec{j})$, we know that $3\vec{i} + \sqrt{3}\vec{j}$ and $\sqrt{3}\vec{i} + \vec{j}$ are scalar multiples of one another, and therefore parallel.

    Since $(\sqrt{3}\vec{i} + \vec{j}) \cdot (\vec{i} - \sqrt{3}\vec{j}) = \sqrt{3} - \sqrt{3} = 0$, we know that $\sqrt{3}\vec{i} + \vec{j}$ and $\vec{i} - \sqrt{3}\vec{j}$ are perpendicular.

    Since $3\vec{i} + \sqrt{3}\vec{j}$ and $\sqrt{3}\vec{i} + \vec{j}$ are parallel, $3\vec{i} + \sqrt{3}\vec{j}$ and $\vec{i} - \sqrt{3}\vec{j}$ are perpendicular, too.

5. We have:

    $$\vec{u} \cdot (\vec{v} + \vec{w}) = (u_1, u_2) \cdot (\vec{v}_1 + \vec{w}_1, \vec{v}_2 + \vec{w}_2)$$
    $$= u_1(v_1 + w_1) + u_2(v_2 + w_2) = u_1 v_1 + u_1 w_1 + u_2 v_2 + u_2 w_2$$
    $$= (u_1 v_1 + u_2 v_2) + (u_1 w_1 + u_2 w_2) = \vec{u} \cdot \vec{v} + \vec{u} \cdot \vec{w}.$$

9. $\vec{c} \cdot \vec{y} = (\vec{i} + 6\vec{j}) \cdot (4\vec{i} - 7\vec{j}) = (1)(4) + (6)(-7) = 4 - 42 = -38$

13. Since $\vec{a} \cdot \vec{y}$ and $\vec{c} \cdot \vec{z}$ are both scalars, the answer to this equation is the product of two numbers and therefore a number. We have

    $$\vec{a} \cdot \vec{y} = (2\vec{j} + \vec{k}) \cdot (4\vec{i} - 7\vec{j}) = 0(4) + 2(-7) + 1(0) = -14$$

    $$\vec{c} \cdot \vec{z} = (\vec{i} + 6\vec{j}) \cdot (\vec{i} - 3\vec{j} - \vec{k}) = 1(1) + 6(-3) + 0(-1) = -17.$$

    Thus,

    $$(\vec{a} \cdot \vec{y})(\vec{c} \cdot \vec{z}) = 238.$$

17. (a) The points A, B and C are shown in Figure 8.7.

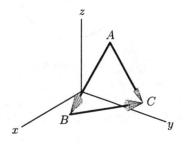

Figure 8.7

First, we calculate the vectors which form the sides of this triangle:

$$\overrightarrow{AB} = (4\vec{i} + 2\vec{j} + \vec{k}) - (2\vec{i} + 2\vec{j} + 2\vec{k}) = 2\vec{i} - \vec{k}$$
$$\overrightarrow{BC} = (2\vec{i} + 3\vec{j} + \vec{k}) - (4\vec{i} + 2\vec{j} + \vec{k}) = -2\vec{i} + \vec{j}$$
$$\overrightarrow{AC} = (2\vec{i} + 3\vec{j} + \vec{k}) - (2\vec{i} + 2\vec{j} + 2\vec{k}) = \vec{j} - \vec{k}.$$

Now we calculate the lengths of each of the sides of the triangles:

$$\|\overrightarrow{AB}\| = \sqrt{2^2 + (-1)^2} = \sqrt{5}$$
$$\|\overrightarrow{BC}\| = \sqrt{(-2)^2 + 1^2} = \sqrt{5}$$
$$\|\overrightarrow{AC}\| = \sqrt{1^2 + (-1)^2} = \sqrt{2}.$$

Thus the length of the shortest side of S is $\sqrt{2}$.

(b) $\cos \angle BAC = \dfrac{\overrightarrow{AB} \cdot \overrightarrow{AC}}{\|\overrightarrow{AB}\| \cdot \|\overrightarrow{AC}\|} = \dfrac{2 \cdot 0 + 0 \cdot 1 + (-1) \cdot (-1)}{\sqrt{5} \cdot \sqrt{2}} \approx 0.32$

Solutions for Section 8.5

1. (a)

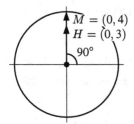

Figure 8.8

Figure 8.8 shows that at 12 noon, we have:
In Cartesian coordinates, $H = (0, 3)$. In polar coordinates, $H = (3, \pi/2)$; that is $r = 3, \theta = \pi/2$. In Cartesian coordinates, $M = (0, 4)$. In polars coordinates, $M = (4, \pi/2)$, that is $r = 4, \theta = \pi/2$.

(b)

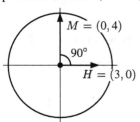

Figure 8.9

Figure 8.9 shows that at 3 pm, we have:
In Cartesian coordinates, $H = (3, 0)$. In polar coordinates, $H = (3, 0)$; that is $r = 3, \theta = 0$. In Cartesians coordinates, $M = (0, 4)$. In polars coordinates, $M = (4, \pi/2)$, that is $r = 4, \theta = \pi/2$.

(c)

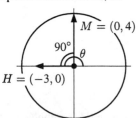

Figure 8.10

Figure 8.10 shows that at 9 am, we have:
In Cartesian coordinates, $H = (-3, 0)$. In Polar, $H = (3, \pi)$; that is $r = 3, \theta = \pi$. In Cartesian coordinates, $M = (0, 4)$. In polars, $M = (4, \pi/2)$, that is $r = 4, \theta = \pi/2$.

(d)

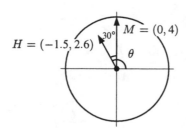

Figure 8.11

Figure 8.11 shows that at 11 am the polar coordinates of the point H are $r = 3$ and $\theta = 30° + 90° = 120° = 2\pi/3$. Thus, the Cartesian coordinates of H are given by

$$x = 3\cos\left(\frac{2\pi}{3}\right) = -\frac{3}{2} = -1.5, \quad y = 3\sin\left(\frac{2\pi}{3}\right) = \frac{3\sqrt{3}}{2} = 2.60.$$

Thus, In Cartesian coordinates, $H = (-3/2, 3\sqrt{3}/2)$. In polar coordinates, $H = (3, 2\pi/3)$. In Cartesian coordinates, $M = (0, 4)$. In polar coordinates, $M = (4, \pi/2)$.

(e)

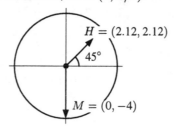

Figure 8.12

Figure 8.12 shows that at 1:30 pm, the polar coordinates of the point H (halfway between 1 and 2 on the clock face) are $r = 3$ and $\theta = 45° = \pi/4$. Thus, the Cartesian coordinates of H are given by

$$x = 3\cos\left(\frac{\pi}{4}\right) = 2.12, \quad y = 3\sin\left(\frac{\pi}{4}\right) = 2.12.$$

we have:
In Cartesian coordinates, $H = (2.12, 2.12)$. In polar coordinates, $H = (3, \pi/4)$. In Cartesian coordinates, $M = (0, -4)$. In polar coordinates, $M = (4, 3\pi/2)$.

(f)

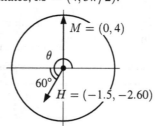

Figure 8.13

Figure 8.13 shows that at 7 am the polar coordinates of the point H are $r = 3$ and $\theta = 60° + 180° = 240° = 4\pi/3$. Thus, the Cartesian coordinates of H are given by

$$x = 3\cos\left(\frac{4\pi}{3}\right) = -\frac{3}{2} = -1.5, \quad y = 3\sin\left(\frac{4\pi}{3}\right) = -\frac{3\sqrt{3}}{2} = -2.60.$$

Thus, In Cartesian coordinates, $H = (-1.5, -2.60)$. In polar coordinates, $H = (3, 4\pi/3)$. In Cartesian coordinates, $M = (0, 4)$. In polar coordinates, $M = (4, \pi/2)$.

(g)

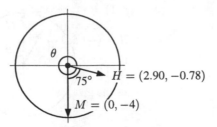

Figure 8.14

Figure 8.14 shows that at 3:30 pm, the polar coordinates of the point H (halfway between 3 and 4 on the clock face) are $r = 3$ and $\theta = 75° + 270° = 23\pi/12$. Thus, the Cartesian coordinates of H are given by

$$x = 3\cos\left(\frac{23\pi}{12}\right) = 2.90, \quad y = 3\sin\left(\frac{23\pi}{12}\right) = -0.78.$$

We have:

In Cartesian coordinates, $H = (2.90, -0.78)$; $M = (0, -4)$.
In polar coordinates, $H = (3, 23\pi/12)$; $M = (4, 3\pi/2)$.

(h)

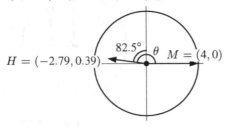

Figure 8.15

Figure 8.15 shows that at 9:15 am, the polar coordinates of the point H (half-way up from 9 and 9:30 on the clock face) are $r = 3$ and $\theta = 82.5° + 90° = 172.5\pi/180$. Thus, the Cartesian coordinates of H are given by

$$x = 3\cos\left(\frac{172.5\pi}{180}\right) = -2.97, \quad y = 3\sin\left(\frac{172.5\pi}{180}\right) = 0.39.$$

we have:

In Cartesian coordinates, $H = (-2.97, 0.39)$. In polar coordinates, $H = (3, 172.5\pi/180)$. In Cartesian coordinates, $M = (4, 0)$. In polar coordinates, $M = (4, 0)$.

Solutions for Chapter 8 Review

1.

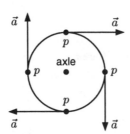

Figure 8.16: $\vec{a}$ is the velocity of P relative to the axle

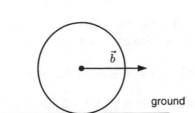

Figure 8.17: $\vec{b}$ is the velocity of the axle relative to the ground

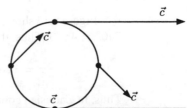

Figure 8.18: $\vec{c}$ is the velocity of P relative to the ground

(a)   Since the wheel is spinning at $6\pi/2\pi(1) = 3$ revolutions/seconds,
       the magnitude $\|\vec{a}\|$ is 3 rev/sec $\cdot 2\pi(1)$ ft/rev $= 6\pi$ ft/sec.

(b) $\|\vec{b}\| = 6\pi$ ft/sec as given.

(c)   If $\vec{c}$ is the velocity of P relative to the ground, then we know that $\vec{c} = \vec{a} + \vec{b}$. We can construct the diagram in Figure 8.18 by adding $\vec{a}$ and $\vec{b}$, which are both of magnitude 6π.

(d)   The point P does stop when it touches the ground. There $\vec{a} = -\vec{b}$ and so $\vec{c} = \vec{a} + \vec{b} = 0$ at that point. The fastest speed is at the top, when $\vec{a} = \vec{b}$ and so $\|\vec{c}\| = \|\vec{a}\| + \|\vec{b}\| = 12\pi$ ft/sec.

5.

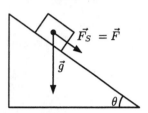

Figure 8.19

If $\theta = 0$ (the plank is at ground level), the sliding force is $F = 0$.
If $\theta = \pi/2$ (the plank is vertical), the sliding force equals g, the force due to gravity.
Therefore, we can guess that F is proportional with $\sin\theta$:

$$F = g\sin\theta.$$

This agrees with the bounds at $\theta = 0$ and $\theta = \pi/2$, and with the fact that the sliding force is smaller than g between 0 and $\pi/2$.

9.   (a)   Clearly $(x-1)^2 + y^2 = 1$ is a circle with center $(1,0)$. To convert this to polar, use $x = r\cos\theta$ and $y = r\sin\theta$. Then $(r\cos\theta - 1)^2 + (r\sin\theta)^2 = 1$ or $r^2\cos^2\theta - 2r\cos\theta + 1 + r^2\sin^2\theta = 1$. This means $r^2(\cos^2\theta + \sin^2\theta) = 2r\cos\theta$, or $r = 2\cos\theta$.

      (b)   12 o'clock $\rightarrow (x,y) = (1,1)$ and $(r,\theta) = (\sqrt{2}, \pi/4)$,
             3 o'clock $\rightarrow (x,y) = (2,0)$ and $(r,\theta) = (2,0)$,
             6 o'clock $\rightarrow (x,y) = (1,-1)$ and $(r,\theta) = (\sqrt{2}, -\pi/4)$,
             9 o'clock $\rightarrow (x,y) = (0,0)$ and $(r,\theta) = (0, \text{any angle })$.
             2

13. $\vec{u} = \vec{i} + \vec{j} + 2\vec{k}$ and $\vec{v} = -\vec{i} + 2\vec{k}$

17.   (a)   To get from A to B, you must go down 7, to the left 2, and forward 2. So $\vec{AB} = 2\vec{i} - 2\vec{j} - 7\vec{k}$. Similarly, $\vec{AC} = -2\vec{i} + 2\vec{j} - 7\vec{k}$.

      (b)   Remember

$$\cos\theta = \frac{\vec{AB} \cdot \vec{AC}}{\|\vec{AB}\|\|\vec{AC}\|} = \frac{(2)(-2) + (-2)(2) + (-7)(-7)}{\sqrt{57}\sqrt{57}} = \frac{41}{57}.$$

So $\theta = 44.00°$.

CHAPTER NINE

1. Yes, $a = 2$, ratio $= 1/2$.

5. Yes, $a = 1$, ratio $= -x$.

9. Sum $= \dfrac{1}{1 - (-x)} = \dfrac{1}{1 + x}, |x| < 1$.

13.

$$\sum_{i=4}^{\infty} \left(\frac{1}{3}\right)^i = \left(\frac{1}{3}\right)^4 + \left(\frac{1}{3}\right)^5 + \cdots = \left(\frac{1}{3}\right)^4 \left(1 + \frac{1}{3} + \left(\frac{1}{3}\right)^2 + \cdots\right) = \frac{(\frac{1}{3})^4}{1 - \frac{1}{3}} = \frac{1}{54}$$

17. Let Q_n represent the quantity, in milligrams, of ampicillin in the blood right after the nth tablet. Then

$$Q_1 = 250$$
$$Q_2 = 250 + 250(0.04)$$
$$Q_3 = 250 + 250(0.04) + 250(0.04)^2$$
$$\vdots$$
$$Q_n = 250 + 250(1.04) + 250(1.04)^2 + \cdots + 250(0.04)^{n-1}.$$

This is a geometric series. Its sum is given by

$$Q_n = \frac{250(1 - (0.04)^n)}{1 - 0.04}.$$

Thus,

$$Q_3 = \frac{250(1 - (0.04)^3)}{1 - 0.04} = 260.40$$

and

$$Q_{40} = \frac{250(1 - (0.04)^{40})}{1 - 0.04} = 260.417.$$

In the long run, as $n \to \infty$, we know that $(0.04)^n \to 0$, and so

$$Q_n = \frac{250(1 - (0.04)^n)}{1 - 0.04} \to \frac{250(1 - 0)}{1 - 0.04} = 260.417.$$

21.

Total present value, in dollars $= 1000 + 1000e^{-0.04} + 1000e^{-0.04(2)} + 1000e^{-0.04(3)} + \cdots$
$$= 1000 + 1000(e^{-0.04}) + 1000(e^{-0.04})^2 + 1000(e^{-0.04})^3 + \cdots$$

This is an infinite geometric series with $a = 1000$ and $x = e^{(-0.04)}$, and sum

Total present value, in dollars $= \dfrac{1000}{1 - e^{-0.04}} = 25{,}503.$

Solutions for Section 9.2

1. Between times $t = 0$ and $t = 1$, x goes at a constant rate from 0 to 1 and y goes at a constant rate from 1 to
    0. So the particle moves in a straight line from $(0, 1)$ to $(1, 0)$. Similarly, between times $t = 1$ and $t = 2$, it
    goes in a straight line to $(0, -1)$, then to $(-1, 0)$, then back to $(0, 1)$. So it traces out the diamond shown in
    Figure 9.1.

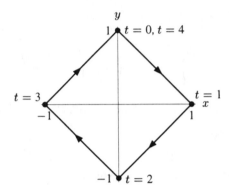

Figure 9.1

5. The particle moves clockwise: For $0 \leq t \leq \frac{\pi}{2}$, we have $x = \cos t$ decreasing and $y = -\sin t$ decreasing.
    Similarly, for the time intervals $\frac{\pi}{2} \leq t \leq \pi, \pi \leq t \leq \frac{3\pi}{2}$, and $\frac{3\pi}{2} \leq t \leq 2\pi$, we see that the particle moves
    clockwise.

9. In all three cases, $y = x^2$, so that the motion takes place on the parabola $y = x^2$.
    In case (a), the x-coordinate always increases at a constant rate of one unit distance per unit time, so the
    equations describe a particle moving to the right on the parabola at constant horizontal speed.
    In case (b), the x-coordinate is never negative, so the particle is confined to the right half of the parabola.
    As t moves from $-\infty$ to $+\infty$, $x = t^2$ goes from ∞ to 0 to ∞. Thus the particle first comes down the right
    half of the parabola, reaching the origin $(0, 0)$ at time $t = 0$, where it reverses direction and goes back up the
    right half of the parabola.
    In case (c), as in case (a), the particle traces out the entire parabola $y = x^2$ from left to right. The
    difference is that the horizontal speed is not constant. This is because a unit change in t causes larger and
    larger changes in $x = t^3$ as t approaches $-\infty$ or ∞. The horizontal motion of the particle is faster when it is
    farther from the origin.

13. The slope of the line is

$$m = \frac{3 - (-1)}{1 - 2} = -4.$$

    The equation of the line with slope -4 through the point $(2, -1)$ is $y - (-1) = (-4)(x - 2)$, so one possible
    parameterization is $x = t$ and $y = -4t + 8 - 1 = -4t + 7$.

17. This curve never closes on itself. Figure 9.2 shows how it starts out.
    The plot for $0 \leq t \leq 8\pi$ is in Figure 9.2.

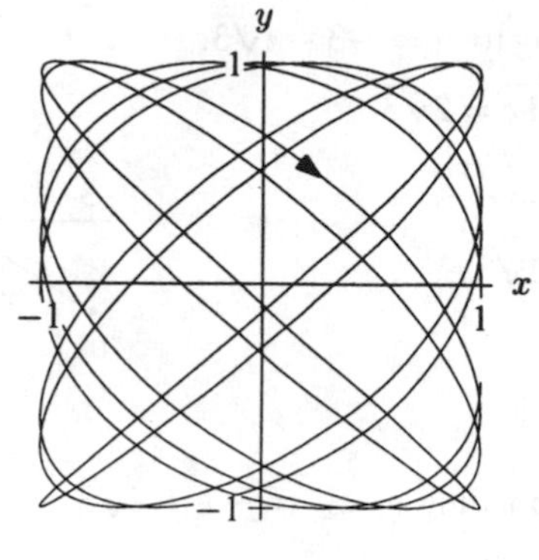

Figure 9.2

Solutions for Section 9.3

1. One possible answer is $x = 3\cos t, y = -3\sin t, 0 \le t \le 2\pi$.

5. The parameterization $(x, y) = (4 + 4\cos t, 4 + 4\sin t)$ gives the correct circle, but starts at $(8, 4)$. To start on the x-axis we need $y = 0$ at $t = 0$, thus

$$(x, y) = \left(4 + 4\cos\left(t - \frac{\pi}{2} \right), 4 + 4\sin\left(t - \frac{\pi}{2} \right) \right).$$

9. The circle $(x - 2)^2 + (y - 2)^2 = 1$.

13. Implicit: $x^2 - 2x + y^2 = 0$, $y < 0$. Explicit: $y = -\sqrt{-x^2 + 2x}, 0 \le x \le 2$. Parametric: The curve is the lower half of a circle centered at $(1, 0)$ with radius 1, so $x = 1 + \cos t, y = \sin t$, for $\pi \le t \le 2\pi$.

Solutions for Section 9.4

1. $2e^{\frac{i\pi}{2}}$

5. $e^{\frac{i3\pi}{2}}$

9. $-5 + 12i$

13. We have $\sqrt{e^{i\pi/3}} = e^{(i\pi/3)/2} = e^{i\pi/6}$, thus $\cos\frac{\pi}{6} + i\sin\frac{\pi}{6} = \frac{\sqrt{3}}{2} + \frac{i}{2}$.

17. One value of $\sqrt[3]{i}$ is $\sqrt[3]{e^{i\frac{\pi}{2}}} = (e^{i\frac{\pi}{2}})^{\frac{1}{3}} = e^{i\frac{\pi}{6}} = \cos\frac{\pi}{6} + i\sin\frac{\pi}{6} = \frac{\sqrt{3}}{2} + \frac{i}{2}$

21. One value of $(\sqrt{3} + i)^{-1/2}$ is
    $(2e^{i\frac{\pi}{6}})^{-1/2} = \frac{1}{\sqrt{2}}e^{i(-\frac{\pi}{12})} = \frac{1}{\sqrt{2}}\cos(-\frac{\pi}{12}) + i\frac{1}{\sqrt{2}}\sin(-\frac{\pi}{12}) \approx 0.683 - 0.183i$

25. (a) $z_1 z_2 = (-3 - i\sqrt{3})(-1 + i\sqrt{3}) = 3 + (\sqrt{3})^2 + i(\sqrt{3} - 3\sqrt{3}) = 6 - i2\sqrt{3}$.

$$\frac{z_1}{z_2} = \frac{-3 - i\sqrt{3}}{-1 + i\sqrt{3}} \cdot \frac{-1 - i\sqrt{3}}{-1 - i\sqrt{3}} = \frac{3 - (\sqrt{3})^2 + i(\sqrt{3} + 3\sqrt{3})}{(-1)^2 + (\sqrt{3})^2} = \frac{i \cdot 4\sqrt{3}}{4} = i\sqrt{3}.$$

(b)   We find (r_1, θ_1) corresponding to $z_1 = -3 - i\sqrt{3}$.

$r_1 = \sqrt{(-3)^2 + (\sqrt{3})^2} = \sqrt{12} = 2\sqrt{3}$.

$\tan \theta_1 = \dfrac{-\sqrt{3}}{-3} = \dfrac{\sqrt{3}}{3}$, so $\theta_1 = \dfrac{7\pi}{6}$.

Thus $-3 - i\sqrt{3} = r_1 e^{i\theta_1} = 2\sqrt{3}\, e^{i\frac{7\pi}{6}}$.

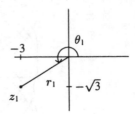

We find (r_2, θ_2) corresponding to $z_2 = -1 + i\sqrt{3}$.

$r_2 = \sqrt{(-1)^2 + (\sqrt{3})^2} = 2$;

$\tan \theta_2 = \dfrac{\sqrt{3}}{-1} = -\sqrt{3}$, so $\theta_2 = \dfrac{2\pi}{3}$.

Thus, $-1 + i\sqrt{3} = r_2 e^{i\theta_2} = 2e^{i\frac{2\pi}{3}}$.

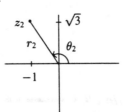

We now calculate $z_1 z_2$ and $\dfrac{z_1}{z_2}$.

$$z_1 z_2 = \left(2\sqrt{3}e^{i\frac{7\pi}{6}}\right)\left(2e^{i\frac{2\pi}{3}}\right) = 4\sqrt{3}e^{i\left(\frac{7\pi}{6} + \frac{2\pi}{3}\right)} = 4\sqrt{3}e^{i\frac{11\pi}{6}}$$

$$= 4\sqrt{3}\left[\cos\frac{11\pi}{6} + i\sin\frac{11\pi}{6}\right] = 4\sqrt{3}\left[\frac{\sqrt{3}}{2} - i\frac{1}{2}\right] = 6 - i2\sqrt{3}.$$

$$\frac{z_1}{z_2} = \frac{2\sqrt{3}e^{i\frac{7\pi}{6}}}{2e^{i\frac{2\pi}{3}}} = \sqrt{3}e^{i\left(\frac{7\pi}{6} - \frac{2\pi}{3}\right)} = \sqrt{3}e^{i\frac{\pi}{2}}$$

$$= \sqrt{3}\left(\cos\frac{\pi}{2} + i\sin\frac{\pi}{2}\right) = i\sqrt{3}.$$

These agree with the values found in (a).

29.   False, since $(1 + i)^2 = 2i$ is not real.

33.   Using Euler's formula, we have:

$$e^{i(2\theta)} = \cos 2\theta + i\sin 2\theta$$

On the other hand,

$$e^{i(2\theta)} = \left(e^{i\theta}\right)^2 = (\cos\theta + i\sin\theta)^2 = (\cos^2\theta - \sin^2\theta) + i(2\cos\theta\sin\theta)$$

Equating imaginary parts, we find

$$\sin 2\theta = 2\sin\theta\cos\theta.$$

Solutions for Section 9.5

1. (a) $\sin(15° + 42°) = \sin 15 \cos 42 + \sin 42 \cos 15 = 0.839$.
   (b) $\sin(15° - 42°) = \sin 15 \cos 42 - \sin 42 \cos 15 = -0.454$.

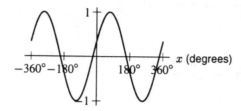

Figure 9.3

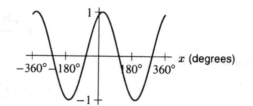

Figure 9.4

(c) $\cos(15° + 42°) = \cos 15 \cos 42 - \sin 15 \sin 42 = 0.544$.
(d) $\cos(15° - 42°) = \cos 15 \cos 42 + \sin 15 \sin 42 = 0.891$.

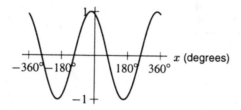

Figure 9.5

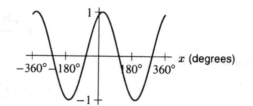

Figure 9.6

5. We can use the product-to-sum identity

$$\sin \alpha \sin \beta = \frac{1}{2} \left[\cos(\alpha - \beta) - \cos(\alpha + \beta) \right]$$

to reduce the right-hand side by putting $\alpha = (u + v)/2$ and $\beta = (u - v)/2$:

$$-2 \sin \left(\frac{u+v}{2} \right) \sin \left(\frac{u-v}{2} \right) = -2 \left[\frac{1}{2} \cos \left(\frac{u+v}{2} - \frac{u-v}{2} \right) - \cos \left(\frac{u+v}{2} + \frac{u-v}{2} \right) \right]$$

$$= -1 \cos \left(\frac{2v}{2} \right) - \cos \left(\frac{2u}{u} \right)$$

$$= \cos u - \cos v.$$

Quite Easily Done.

9.

$$\cos 3\theta = \cos(2\theta + \theta) = \cos 2\theta \cos \theta + \sin 2\theta \sin \theta$$

$$= (2 \cos^2(\theta) - 1) \cos \theta - (2 \sin \theta \cos \theta) \sin \theta$$

$$= 2 \cos^3 \theta - \cos \theta - 2 \cos \theta (\sin^2 \theta)$$

$$= 2 \cos^3 \theta - \cos \theta - 2 \cos \theta (1 - \cos^2 \theta)$$

$$= 4 \cos^3 \theta - 3 \cos \theta$$

Solutions for Section 9.6

1. Substitute $x = 0$ into the formula for $\sinh x$. This yields
$$\sinh 0 = \frac{e^0 - e^{-0}}{2} = \frac{1 - 1}{2} = 0.$$

5. First, we observe that
$$\cosh 2x = \frac{e^{2x} + e^{-2x}}{2}.$$

Now, using the fact that $e^x \cdot e^{-x} = 1$, we calculate

$$\begin{aligned}
\cosh^2 x &= \left(\frac{e^x + e^{-x}}{2}\right)^2 \\
&= \frac{(e^x)^2 + 2e^x \cdot e^{-x} + (e^{-x})^2}{4} \\
&= \frac{e^{2x} + 2 + e^{-2x}}{4}.
\end{aligned}$$

Similarly, we have

$$\begin{aligned}
\sinh^2 x &= \left(\frac{e^x - e^{-x}}{2}\right)^2 \\
&= \frac{(e^x)^2 - 2e^x \cdot e^{-x} + (e^{-x})^2}{4} \\
&= \frac{e^{2x} - 2 + e^{-2x}}{4}.
\end{aligned}$$

Thus, to obtain $\cosh 2x$, we need to add (rather than subtract) $\cosh^2 x$ and $\sinh^2 x$, giving

$$\begin{aligned}
\cosh^2 x + \sinh^2 x &= \frac{e^{2x} + 2 + e^{-2x} + e^{2x} - 2 + e^{-2x}}{4} \\
&= \frac{2e^{2x} + 2e^{-2x}}{4} \\
&= \frac{e^{2x} + e^{-2x}}{2} \\
&= \cosh 2x.
\end{aligned}$$

Thus, we see that the identity relating $\cosh 2x$ to $\cosh x$ and $\sinh x$ is
$$\cosh 2x = \cosh^2 x + \sinh^2 x.$$

Solutions for Chapter 9 Review

1. Yes, $a = 1$, ratio $= -y^2$.

5. (a)   Let h_n be the height of the nth bounce after the ball hits the floor for the nth time. Then from Figure 9.7,

$$h_0 = \text{height before first bounce} = 10 \text{ feet,}$$

$$h_1 = \text{height after first bounce} = 10 \left(\frac{3}{4}\right) \text{ feet,}$$

$$h_2 = \text{height after second bounce} = 10 \left(\frac{3}{4}\right)^2 \text{ feet.}$$

Generalizing this gives

$$h_n = 10 \left(\frac{3}{4}\right)^n.$$

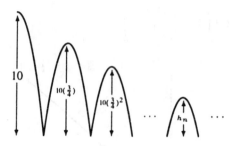

Figure 9.7

(b)   When the ball hits the floor for the first time, the total distance it has traveled is just $D_1 = 10$ feet. (Notice that this is the same as $h_0 = 10$.) Then the ball bounces back to a height of $h_1 = 10 \left(\frac{3}{4}\right)$, comes down and hits the floor for the second time. The total distance it has traveled is

$$D_2 = h_0 + 2h_1 = 10 + 2 \cdot 10 \left(\frac{3}{4}\right) = 25 \text{ feet.}$$

Then the ball bounces back to a height of $h_2 = 10 \left(\frac{3}{4}\right)^2$, comes down and hits the floor for the third time. It has traveled

$$D_3 = h_0 + 2h_1 + 2h_2 = 10 + 2 \cdot 10 \left(\frac{3}{4}\right) + 2 \cdot 10 \left(\frac{3}{4}\right)^2 = 25 + 2 \cdot 10 \left(\frac{3}{4}\right)^2 = 36.25 \text{ feet.}$$

Similarly,

$$
\begin{aligned}
D_4 &= h_0 + 2h_1 + 2h_2 + 2h_3 \\
&= 10 + 2 \cdot 10 \left(\frac{3}{4}\right) + 2 \cdot 10 \left(\frac{3}{4}\right)^2 + 2 \cdot 10 \left(\frac{3}{4}\right)^3 \\
&= 36.25 + 2 \cdot 10 \left(\frac{3}{4}\right)^3 \\
&\approx 44.69 \text{ feet.}
\end{aligned}
$$

(c) When the ball hits the floor for the nth time, its last bounce was of height h_{n-1}. Thus, by the method used in part (b), we get

$$D_n = h_0 + 2h_1 + 2h_2 + 2h_3 + \cdots + 2h_{n-1}$$

$$= 10 + 2 \cdot 10 \left(\frac{3}{4}\right) + 2 \cdot 10 \left(\frac{3}{4}\right)^2 + 2 \cdot 10 \left(\frac{3}{4}\right)^3 + \cdots + 2 \cdot 10 \left(\frac{3}{4}\right)^{n-1}$$

$$\underbrace{\phantom{= 10 + 2 \cdot 10 \left(\frac{3}{4}\right) + 2 \cdot 10 \left(\frac{3}{4}\right)^2 + 2 \cdot 10 \left(\frac{3}{4}\right)^3 + \cdots + 2 \cdot 10 \left(\frac{3}{4}\right)^{n-1}}}_{\text{finite geometric series}}$$

$$= 10 + 2 \cdot 10 \cdot \left(\frac{3}{4}\right)\left(1 + \left(\frac{3}{4}\right) + \left(\frac{3}{4}\right)^2 + \cdots + \left(\frac{3}{4}\right)^{n-2}\right)$$

$$= 10 + 15 \left(\frac{1 - \left(\frac{3}{4}\right)^{n-1}}{1 - \left(\frac{3}{4}\right)}\right)$$

$$= 10 + 60 \left(1 - \left(\frac{3}{4}\right)^{n-1}\right).$$

9.

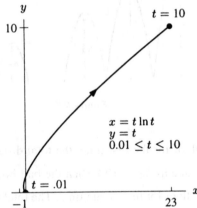

Figure 9.8

After a short move to the left, the particle moves steadily to the right.

13. The plot looks like Figure 9.9.

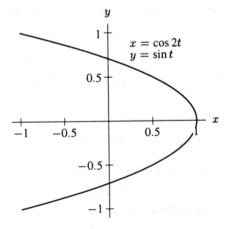

Figure 9.9

which does appear to be part of a parabola. To prove that it is, we note that we have

$$x = \cos 2t$$

$$y = \sin t$$

and must somehow find a relationship between x and y. Recall the trigonometric identity

$$\cos 2t = 1 - 2\sin^2 t.$$

Thus we have $x = 1 - 2y^2$, which is a parabola lying along the x-axis, for $-1 \le y \le 1$.

17. $13e^{i\theta}$, where $\theta = \arctan(-\frac{12}{5}) \approx -1.176$ is an angle in the fourth quadrant.

21. One value of $(-4 + 4i)^{2/3}$ is $[\sqrt{32}e^{i\frac{3\pi}{4}}]^{2/3} = (\sqrt{32})^{2/3}e^{i\frac{\pi}{2}} = 2^{\frac{10}{3}}\cos\frac{\pi}{2} + i2^{\frac{10}{3}}\sin\frac{\pi}{2} = 8i\sqrt[3]{2}$

25. (a) From $\cos 2u = 2\cos^2 u - 1$, we obtain $\cos u = \pm\sqrt{\dfrac{1 + \cos 2u}{2}}$ and letting $u = \frac{v}{2}$, $\cos\dfrac{v}{2} = \pm\sqrt{\dfrac{1 + \cos v}{2}}$.

  (b) From $\tan\frac{1}{2}v = \dfrac{\sin\frac{1}{2}v}{\cos\frac{1}{2}v} = \dfrac{\pm\sqrt{\dfrac{1 - \cos v}{2}}}{\pm\sqrt{\dfrac{1 + \cos v}{2}}}$ we simplify to get $\tan\frac{1}{2}v = \pm\sqrt{\dfrac{1 - \cos v}{1 + \cos v}}$.

  (c) The sign of $\sin\frac{1}{2}v$ is $+$, the sign of $\cos\frac{1}{2}v$ is $-$, and the sign of $\tan\frac{1}{2}v$ is $-$.

  (d) The sign of $\sin\frac{1}{2}v$ is $-$, the sign of $\cos\frac{1}{2}v$ is $-$, and the sign of $\tan\frac{1}{2}v$ is $+$.

  (e) The sign of $\sin\frac{1}{2}v$ is $-$, the sign of $\cos\frac{1}{2}v$ is $+$, and the sign of $\tan\frac{1}{2}v$ is $-$.

APPENDIX

Solutions for Section A

1. Since $\dfrac{1}{7^{-2}}$ is the same as y^2, we obtain $7 \cdot 7$ or 49.

5. The order of operations tells us to square 3 first (giving 9) and then multiply by -2. Therefore $(-2)\left(3^2\right) = (-2)(9) = -18$.

9. First we see within the radical that $(-4)^2 = 16$. Therefore $\sqrt{(-4)^2} = \sqrt{16} = 4$.

13. For this example, we have $\left(\dfrac{1}{27}\right)^{-1/3} = (27)^{1/3} = 3$. This is because $\left(\dfrac{1}{27}\right)^{-1/3} = \left(\left(\dfrac{1}{27}\right)^{-1}\right)^3 = \left(\dfrac{27}{1}\right)^{1/3} = 3$.

17. If we expand $\left(4L^{2/3}P\right)^{3/2}$, we obtain $4^{3/2} \cdot L^1 \cdot P^{3/2}$ and then multiplying by $P^{-3/2}$ yields $\left(4^{3/2} \cdot L^1 \cdot P^{3/2}\right)P^{-3/2} = 8LP^0 = 8L$.

21.
$$(3AB)^{-1}\left(A^2B^{-1}\right)^2 = \left(3^{-1} \cdot A^{-1} \cdot B^{-1}\right)\left(A^4 \cdot B^{-2}\right) = \frac{A^4}{3^1 \cdot A^1 \cdot B^1 \cdot B^2} = \frac{A^3}{3B^3}.$$

25. $x^e\left(x^e\right)^2 = x^e \cdot x^{2e} = x^{e+2e} = x^{3e}$

29. $\dfrac{a^{n+1}3^{n+1}}{a^n 3^n} = a^{n+1-n}3^{n+1-n} = a^1 \cdot 3^1 = 3a$

33. $-625^{3/4} = -(\sqrt[4]{625})^3 = -(5)^3 = -125$

37. $-64^{3/2} = -(\sqrt{64})^3 = -(8)^3 = -512$

Solutions for Section B

1. First we multiply 4 by the terms $3x$ and $-2x^2$, and then use foil to expand $(5 + 4x)(3x - 4)$. Therefore,
$$\left(3x - 2x^2\right)(4) + (5 + 4x)(3x - 4) = 12x - 8x^2 + 15x - 20 + 12x^2 - 16x$$
$$= 4x^2 + 11x - 20.$$

5. First we square $\sqrt{2x} + 1$ and then take the negative of this result. Therefore,
$$-\left(\sqrt{2x} + 1\right)^2 = -\left(\sqrt{2x} + 1\right)\left(\sqrt{2x} + 1\right) = -\left(2x + \sqrt{2x} + \sqrt{2x} + 1\right)$$
$$= -(2x + 2\sqrt{2x} + 1) = -2x - 2\sqrt{2x} - 1.$$

9. Using foil we obtain:
$$(x + 3)\left(\frac{24}{x} + 2\right) = 24 + 2x + \frac{72}{x} + 6$$
$$= 30 + 2x + \frac{72}{x}.$$

Solutions for Section C

1. Since each term has a common factor of 2, we write:

$$2x^2 - 10x + 12 = 2\left(x^2 - 5x + 6\right)$$
$$= 2(x - 3)(x - 2).$$

5. The expression $x^2 + y^2$ cannot be factored.

9. The idea here is to rewrite the second expression $-2(s - r)$ as $+2(r - s)$. This latter expression shares a comon factor of $r - s$ with the first expression $r(r - s)$. Thus,

$$r(r - s) - 2(s - r) = r(r - s) + 2(r - s) = (r + 2)(r - s).$$

13. Factor as:

$$e^{2x} + 2e^x + 1 = (e^x + 1)(e^x + 1) = (e^x + 1)^2.$$

Solutions for Section D

1. The common denominator is $(x - 4)(x + 4) = x^2 - 16$. Therefore,

$$\frac{3}{x - 4} - \frac{2}{x + 4} = \frac{3(x + 4)}{(x - 4)(x + 4)} - \frac{2(x - 4)}{(x + 4)(x - 4)}$$
$$\frac{3(x + 4) - 2(x - 4)}{x^2 - 16} = \frac{3x + 12 - 2x + 8}{x^2 - 16}$$
$$= \frac{x + 20}{x^2 - 16}.$$

5. The common denominator is $(\sqrt{x})^3$.

$$\frac{1}{\sqrt{x}} - \frac{1}{(\sqrt{x})^3} = \frac{(\sqrt{x})^2}{(\sqrt{x})^3} - \frac{1}{(\sqrt{x})^3} = \frac{x - 1}{(\sqrt{x})^3}$$

It is fine to leave the answer in the form $\frac{x-1}{(\sqrt{x})^3}$, or we can rationalize the denominator:

$$\frac{x - 1}{(\sqrt{x})^3} = \frac{x - 1}{x\sqrt{x}} = \frac{\sqrt{x}(x - 1)}{x\sqrt{x}\sqrt{x}} = \frac{x\sqrt{x} - \sqrt{x}}{x^2}.$$

9. Each of the denominators are different and therefore the common denominator is $r_1 r_2 r_3$. Accordingly,

$$\frac{1}{r_1} + \frac{1}{r_2} + \frac{1}{r_3} = \frac{r_2 r_3 + r_1 r_3 + r_1 r_2}{r_1 r_2 r_3}.$$

13. Recall that the terms a^{-2} and b^{-2} can be written as $\frac{1}{a^2}$ and $\frac{1}{b^2}$ respectively. Therefore,

$$\frac{a^{-2} + b^{-2}}{a^2 + b^2} = \frac{\frac{1}{a^2} + \frac{1}{b^2}}{a^2 + b^2} = \frac{\frac{b^2 + a^2}{a^2 b^2}}{a^2 + b^2} = \frac{b^2 + a^2}{a^2 b^2} \cdot \frac{1}{a^2 + b^2} = \frac{1}{a^2 b^2}.$$

17. We cancel the common factor $x^3 + 1$ in both numerator and denominator. Therefore,

$$\frac{2x(x^3 + 1)^2 - x^2(2)(x^3 + 1)(3x^2)}{[(x^3 + 1)^2]^2} = \frac{2x(x^3 + 1) - x^2(2)(3x^2)}{(x^3 + 1)^3}$$

$$= \frac{2x^4 + 2x - 6x^4}{(x^3 + 1)^3} = \frac{2x - 4x^4}{(x^3 + 1)^3}.$$

21. Dividing $2x^3$ into each term in the numerator yields:

$$\frac{26x + 1}{2x^3} = frac26x2x^3 + \frac{1}{2x^3} = \frac{13}{x^2} + \frac{1}{2x^3}.$$

25.

$$\frac{\frac{1}{3}x - \frac{1}{2}}{2x} = \frac{\frac{x}{3}}{2x} - \frac{\frac{1}{2}}{2x} = \frac{x}{3} \cdot \frac{1}{2x} - \frac{1}{2} \cdot \frac{1}{2x} = \frac{1}{6} - \frac{1}{4x}$$

29. Dividing the denominator R into each term in the numerator yields,

$$\frac{R + 1}{R} = \frac{R}{R} + \frac{1}{R} = 1 + \frac{1}{R}.$$

Solutions for Section E

1. $3x^2\left(x^{-1}\right) + \dfrac{1}{2x} + x^2 + \dfrac{1}{5} = 3x^1 + \dfrac{1}{2}x^{-1} + x^2 + \dfrac{1}{5}$

5.

$$2P^2(P) + (9P)^{1/2} = 2P^3 + 3P^{1/2}$$

9.

$$\frac{-3(4x - x^2)}{7x} = \frac{-12x + 3x^2}{7x} = -\frac{12}{7} + \frac{3}{7}x$$

13. We write $(1 - x)$ as $-(x - 1)$ and then multiply the factor $(x - 1)^3$ by $-(x - 1)$ which both have the same base. Therefore:

$$0.7(x - 1)^3(1 - x) = 0.7(x - 1)^3(-1(x - 1))$$

$$= -0.7(x - 1)^3(x - 1) = -0.7(x - 1)^4.$$

17. Since the numerator and denominator are both raised to the same exponent, we write

$$\frac{1^x}{2^x} = \left(\frac{1}{2}\right)^x.$$

21. Recall that $3^{-x} = \left(\dfrac{1}{3}\right)^x$. Therefore

$$2 \cdot 3^{-x} = 2\left(\frac{1}{3}\right)^x.$$

25. Both numerator and denominator bases are raised to the same exponent of x. Therefore

$$\frac{5^x}{-3x} = -\left(\frac{5}{3}\right)^x.$$

29. First we factor out -1. Then

$$-x^2 + 6x - 2 = -(x^2 - 6x + 2) = -(x^2 - 6x + 9 - 9 + 2)$$
$$= -(x^2 - 6x + 9 - 7) = -(x^2 - 6x + 9) + 7$$
$$= -(x - 3)^2 + 7.$$

33. We write

$$-(\sin(\pi t))^{-1}(-\cos(\pi t))\pi = \frac{-(-\cos(\pi t))\pi}{\sin(\pi t)} = \frac{\pi \cos(\pi t)}{\sin(\pi t)}.$$

37.

$$\frac{1}{2}(x^2 + 16)^{-1/2}(2x) = \frac{1}{2}(2x)(x^2 + 16)^{-1/2} = x(x^2 + 16)^{-1/2}$$
$$= \frac{x}{(x^2 + 16)^{1/2}} = \frac{x}{\sqrt{x^2 + 16}}$$

Solutions for Section F

1. We first distribute $\frac{5}{3}(y + 2)$ to obtain:

$$\frac{5}{3}(y + 2) = \frac{1}{2} - y$$
$$\frac{5}{3}y + \frac{10}{3} = -\frac{1}{2} - y$$
$$\frac{5}{3}y + y = \frac{1}{2} - \frac{10}{3}$$
$$\frac{5}{3}y + \frac{3y}{3} = \frac{3}{6} - \frac{20}{6}$$
$$\frac{8y}{3} = -\frac{17}{16}$$
$$\left(\frac{3}{8}\right)\frac{8y}{3} = \left(\frac{3}{8}\right)\left(-\frac{17}{6}\right)$$
$$y = -\frac{17}{16}.$$

5. First multiply both sides by (-1):

$$-1(8 + 2x - 3x^2) = (-1)(0).$$

$$3x^2 - 2x - 8 = 0$$
$$(3x + 4)(x - 2) = 0$$
$$3x + 4 = 0 \quad \text{or} \quad x - 2 = 0$$
$$x = -\frac{4}{3} \quad \text{or} \quad x = 2.$$

9. We write $x^2 - 1 = 2x$ or $x^2 - 2x - 1 = 0$ which doesn't factor. Employing the quadratic formula, we have $a = 1, b = -2, c = -1$. Therefore

$$x = \frac{-(-2) \pm \sqrt{(-2)^2 - 4(1)(-1)}}{2(1)} = \frac{2 \pm \sqrt{4 + 4}}{2} = \frac{2 \pm \sqrt{8}}{2}$$
$$= \frac{2 \pm 2\sqrt{2}}{2} = 1 \pm \sqrt{2}.$$

13. To find the common denominator, we factor the second denominator

$$\frac{2}{z - 3} + \frac{7}{z^2 - 3z} = 0$$

$$\frac{2}{z - 3} + \frac{7}{z(z - 3)} = 0$$

which produces a common denominator of $z(z - 3)$. Therefore:

$$\frac{2z}{z(z - 3)} + \frac{7}{z(z - 3)} = 0$$
$$\frac{2z + 7}{z(z - 3)} = 0$$
$$2z + 7 = 0$$
$$z = -\frac{7}{2}.$$

17. We can solve this equation by squaring both sides.

$$\sqrt{r^2 + 24} = 7$$
$$r^2 + 24 = 49$$
$$r^2 = 25$$
$$r = \pm 5$$

21. First we simplify the equation and then take the natural log of both sides of the equation.

$$5000 = 2500(0.97)^t$$
$$2 = (0.97)^t$$
$$\ln 2 = \ln(0.97)^t$$
$$\ln 2 = t \ln(0.97)$$
$$t = \frac{\ln 2}{\ln(0.97)} \approx -22.76$$

25. We begin by squaring both sides of the equation in order to eliminate the radical.

$$T = 2\pi\sqrt{\frac{l}{g}}$$

$$T^2 = 4\pi^2\left(\frac{l}{g}\right)$$

$$\frac{gT^2}{4\pi^2} = l$$

29.

$$l = l_0 + \frac{k}{2}w$$

$$l - l_0 = \frac{k}{2}w$$

$$2(l - l_0) = kw$$

$$\frac{2}{k}(l - l_0) = w$$

Solutions for Section G

1. We substitute the expression $-\frac{3}{5}x + 6$ for y in the first equation.

$$2x + 3y = 7$$

$$2x + 3\left(-\frac{3}{5}x + 6\right) = 7$$

$$2x - \frac{9}{5}x + 18 = 7 \quad \text{or}$$

$$\frac{10}{5}x - \frac{9}{5}x + 18 = 7$$

$$\frac{1}{5}x + 18 = 7$$

$$\frac{1}{5}x = -11$$

$$x = -55$$

$$y = -\frac{3}{5}(-55) + 6$$

$$y = 39$$

5. We set the equations $y = x$ and $y = 3 - x$ equal to one another.

$$x = 3 - x$$

$$2x = 3$$

$$x = \frac{3}{2} \quad \text{and} \quad y = \frac{3}{2}$$

So the point of intersection is $(3/2, 3/2)$.

Solutions for Section H

1.

$$2(x - 7) \geq 0$$
$$2x - 14 \geq 0$$
$$2x \geq 14$$
$$x \geq 7$$

5. The expression $x + 4$ must be greater than or equal to zero. Therefore

$$x + 4 \geq 0$$
$$x \geq -4.$$

9. If we raise a positive number (such as 2) to any power, the result is always positive. Therefore, the solution is all real numbers.

13. $p \neq 5$

17. (a) The expression $3x^2 + 6x$ is defined for all x. Therefore no value for x makes it undefined.
    (b)

$$3x^2 + 6x = 0$$
$$x^2 + 2x = 0$$
$$x(x + 2) = 0$$
$$x = 0 \text{ or } x = -2.$$

   (c) To solve $3x^2 + 6x > 0$, or $x(x + 2) > 0$, we flag the number line at $x = 0$ and $x = -2$ and note the sign on each interval. Therefore

$$
\begin{array}{ccc}
(-)(-) & (-)(+) & (+)(+) \\
\hline
\text{positive} -2 & \text{negative} & 0 \text{ positive}
\end{array}
$$

$$3x^2 + 6x > 0 \text{ when } x > 0 \text{ or } x < -2.$$

   (d) $3x^2 + 6x < 0$ when $-2 < x < 0$.

21. (a) Since $\frac{1}{3}x^{-2/3} = \frac{1}{3\sqrt[3]{x^2}}$, the fraction is undefined when $x = 0$.

   (b) No value of x makes $\frac{1}{\sqrt[3]{x^2}} = 0$.

   (c) Since all x-values except $x = 0$ make $x^2 > 0$, we have $\frac{1}{3\sqrt[3]{x^2}} > 0$ for $x \neq 0$.

   (d) No value of x makes $\frac{1}{3\sqrt[3]{x^2}} < 0$.

25. (a) The values that make the fraction undefined are the solutions to $(u^2 + 1)^3 = 0$. However, $u^2 \geq 0$, so $u^2 + 1 \geq 1$, giving $(u^2 + 1)^3 \geq 1$. There are no values of x that make the expression undefined.

(b)

$$\frac{1 - 3u^2}{(u^2 + 1)^3} = 0$$

$$1 - 3u^2 = 0$$

$$3u^2 = 1$$

$$u^2 = \frac{1}{3}$$

$$u = +\sqrt{\frac{1}{3}} \text{ or } u = -\sqrt{\frac{1}{3}}$$

$$u = +\frac{1}{\sqrt{3}} \text{ or } u = -\frac{1}{\sqrt{3}}.$$

(c) To solve $\frac{1-3u^2}{(u^2+1)^3} > 0$, flag $u = \pm\frac{1}{\sqrt{3}}$ on the numberline.

$$\frac{(-)}{(+)} \qquad \frac{(+)}{(+)} \qquad \frac{(-)}{(+)}$$

negative $\frac{1}{\sqrt{3}}$ positive $\frac{1}{\sqrt{3}}$ negative

The answer is $-\frac{1}{\sqrt{3}} < u < \frac{1}{\sqrt{3}}$.

(d) $\dfrac{1 - 3u^2}{(u^2 + 1)^3} < 0$ for $u < -\dfrac{1}{\sqrt{3}}, u > \dfrac{1}{\sqrt{3}}$

29. We subtract $\dfrac{1}{2}$ from all terms of the inequality and then divide by -1. This latter step will reverse the inequality.

$$0 \le \frac{1}{2} - n < 11$$

$$-\frac{1}{2} \le -n < \frac{21}{2}$$

$$\frac{1}{2} \ge n > -\frac{21}{2}$$

33.

$$2 + \frac{r}{r - 3} > 0$$

$$\frac{r}{r - 3} > -2$$

If $r - 3 > 0$, then we can multiply both sides by $(r - 3)$ without reversing the inequality. Note that $r - 3 > 0$ gives us the condition $r > 3$. So if $r > 3$, then

$$r > -2(r - 3)$$

$$r > -2r + 6$$

$$3r > 6$$

$$r > 2.$$

But we already have $r > 2$ by requiring $r > 3$. Thus, $r > 3$ is part of the solution.

If $r - 3 < 0$, then we must reverse the inequality when we multiply both sides by $(r - 3)$. So, if $r - 3 < 0$, that is, $r < 3$, then

$$r < -2(r - 3)$$
$$r < -2r + 6$$
$$3r < 6$$
$$r < 2.$$

The final solution is $r > 3$ or $r < 2$.

Notes

Notes

Notes

Notes

Notes

Notes

Notes

Notes

Notes

Notes

Notes

Notes

Notes

Notes